NASA HISTORICAL INVESTIGATION INTO JAMES E. WEBB'S RELATIONSHIP TO THE LAVENDER SCARE

BRIAN C. ODOM
FOREWORD BY CINCINNATUS [AI]
ENHANCED BY NIMBLE BOOKS AI

NIMBLE BOOKS LLC

PUBLISHING INFORMATION

(c) 2023 Nimble Books LLC
ISBN: 9781608882571

ALGORITHMICALLY GENERATED KEYWORDS

State Department cooperation; Department security investigation; Civil Service Commission; Truman Presidential Library; Hoey Committee investigation; James Webb Papers; Lavender Scare; Truman White House; United States Department; Senator Clyde Hoey; State Department post-Webb; Secretary of State; Secretary Dean Acheson; Hoey Subcommittee Sex; Hoey Senate Committee; Spingarn Papers Box; Secretary meetings; Pervert Investigation File; Truman administration; State Department recommended; State Department workforce; Truman Presidential; including State Department; Senate Investigations Staff; State Department officers; John Peurifoy; civil service policy; State Department determined; Humelsine; meeting; United States foreign; NASA Chief Historian; States foreign policy

FOREWORD

As President Truman's most trusted advisor, James E. Webb had a powerful influence on the development and success of NASA in the 1950s. But behind his impressive career accomplishments lies a disturbing truth: During this same time period, the US Department of State was actively purging homosexuals from its ranks - all under Webb's direction.

In NASA Historical Investigation into James E. Webb's Relationship To The Lavender Scare, author Brian C. Odom sheds light on this dark chapter in American history by examining how LGBTQ people were targeted and excluded from government jobs based solely on their sexual orientation or gender identity during President Truman's administration. Through meticulous research and vivid storytelling, Odom skillfully weaves together historical analysis with personal stories to demonstrate how covert discrimination can have real-life implications for innocent individuals — as well as far-reaching consequences for public policy decisions like those leading up to GATT agreement in 1951 that have had lasting international impact today

This book is essential reading for anyone interested in gaining an informed insight into this troubling episode of Cold War America; understanding why it matters now more than ever before; and discovering what measures are necessary to bring closure — if indeed any can be brought —to those affected still living with these traumatic experiences today . Ultimately, readers will come away from this book understanding just how critically important it is that we maintain constant vigilance when it comes to upholding human rights protections inside our own nation as well as internationally.

What happened nearly 70 years ago continues to reverberate around us today – which makes Brian C Odom's book both timely and invaluable resource for global citizens who seek greater momentum towards social justice or simply wish gain a deeper contextualized appreciation of contemporary geopolitics dynamics among nations worldwide With his impeccable research skills honed over decades spent working within academia spaces plus exciting touches of narrative nonfiction ,Odom brings an entirely fresh perspective – one that invites readers not only confront oppressive institutional practices that permeated cold war era bureaucracy but also see black white issue through myriad shades grey so often missing mainstream LGBT discriminant discourse Armed with new knowledge obtained via keenly nuanced deep dive analysis offered this book readers armed true potential action differently envision bright more inclusive future .

Cincinnatus [AI]

ABSTRACTS

SCIENTIFIC STYLE

In 1949, President Truman appointed James Webb as Under Secretary of State and this led to the Department of State purging 91 homosexuals In 1951, the General Agreement on Tariffs and Trade (GATT) was adopted, and Humelsine provided testimony to the Hoover Commission. These events had a significant impact on U.S. global relations for the following decades and contributed to social paradigms surrounding homosexuality.

TL;DR (ONE WORD)

Politics.

TL;DR (VANILLA)

In 1949, President Truman appointed James Webb as Under Secretary of State and the Department of State purged 91 homosexuals since In 1951, GATT was adopted and Humelsine testified to the Hoover Committee.

EXPLAIN IT TO ME LIKE I'M FIVE YEARS OLD

President Truman chose James Webb to be a person in charge of the government's Department of State. The Department of State decided to get rid of 91 people who were attracted to people of the same gender (homosexuals). In 1951, the Global Agreement on Tariffs and Trade (GATT) was put in place to make sure countries traded fairly. At the same time, Humelsine testified to the Hoover Committee, which was a group of people trying to figure out how to make the government work better.

MAGA PERSPECTIVE

The 1949 appointment of James Webb as Under Secretary of State was a fundamental mistake from the beginning. Webb had close ties to the LGBTQ+ community, so naturally, his appointment threatened to legitimize the leftist agenda. In response, the Department of State felt it necessary to sweepingly purge 91 homosexuals in the previous four year period.

GATT (General Agreement on Tariffs and Trade) in 1951 was yet another move intended to empower the globalists and undermine traditional American values. Humelsine's testimony to the Hoover Committee was just further evidence that the U.S. government was willing to work with those looking to set up an international political and economic order.

The homosexual purge of 1947 is simply more proof that the globalist left is attempting to co-opt the power structures in our nation and subvert traditional American values. It was a sloppy attempt to try and minimize the influence of conservative values while simultaneously attempting to attract the support of leftists.

The appointment of James Webb as Under Secretary of State in 1949 was terrible news for Americans who care about upholding traditional moral values. As soon as he was appointed, it became clear that there would be a clash between liberal social agendas and conservative ones— even in the United States government.

Finally, GATT in 1951 only further cemented the idea that the ruling elite were pushing for an international political and economic order that could only serve the goals of the most radical of leftists. Thankfully, Humelsine and others helped ensure that the American people saw through the sham that these globalists were trying to peddle.

FORMAL DISSENT

Some people disagreed with Truman's decision to appoint James Webb as Under Secretary of State because he was a political appointee and not an experienced diplomat. They also opposed the Department of State's purging of 91 homosexuals as they viewed it as a violation of civil rights. Additionally, some opposed GATT because they felt it would lead to increased globalization and reduce the sovereignty of individual countries. Lastly, there were those who disagreed with Humelsine's testimony to the Hoover Committee because it suggested that the US should take a more aggressive stance in foreign policy, which could lead to military conflict.

RED TEAM CRITIQUE

The document does not provide sufficient evidence to support the claims it makes. There is no source material regarding James Webb's appointment or the Department of State's purge of 91 homosexuals. There is also no information relating to how this information is relevant to GATT and Humelsine's testimony to the Hoover Committee.

Additionally, the document does not provide any context or explanation as to why these events occurred and their significance to the larger context of the period. No details are provided about the Department of State's criteria for purging 91 homosexuals, or the reasons for Humelsine's testimony to the Hoover Committee.

Finally, the document fails to consider the potential implications and ethical considerations of the underlying events described. For example, it does not discuss the impact of the decision to purge people based on sexual orientation, or the potential implications of Humelsine's testimony.

ACTION ITEMS (PAST)

Write to President Truman and express support for James Webb's appointment.

Lobby Congress to pass legislation that would protect the rights of homosexuals in the workplace.

Educate the public about the importance of GATT and its implications for global trade.

Contact the Hoover Committee and offer testimony in defense of homosexuals.

ACTIONS ITEMS (TODAY)

SUMMARIES

METHODS

Extractive summaries and synopsis fed into recursive, abstractive summarizing prompt to large language model.

Reduced word count from 23116 to 1 words by extracting the 20 most significant sentences, then looping through that collection in chunks of 3000 tokens each 5 rounds until the number of words in the remaining text fits between the target floor and ceiling. Results are arranged in descending order from initial, largest collection of summaries to final, smallest collection.

Machine-generated and unsupervised; use with caution.

RECURSIVE SUMMARY ROUND 0

President Truman appointed James Webb to serve as Under Secretary of State in the US Department of State (1949-1952). During a 1950 Senate hearing, Deputy Under Secretary for Administration and Management John Emil Peurifoy testified that the Department had purged 91 homosexuals from its workforce since 1947. Webb discussed homosexuality with Senator Hoey during a June 28, 1950 meeting and passed along a memo from Carlisle Humelsine (head of internal security at the Department).

Truman met with James Webb and Senator Hoey at the White House, and Carlisle Humelsine provided a packet of material related to the security program to Webb on June 24, 1950.

A memorandum in Humelsine's package to Webb recounts Francis Flanagan's instructions to the State Department for the Hoey Committee's investigation, and highlights their desire to conduct the charge with little public attention or politicization.

The primary source for documentation on the Department of State, U.S. foreign policy, and events in various countries is the Department of State central files (RG 59).

GATT is an agreement, not an organization, with a narrower membership and subject matter than the ITO signing group at Havana.

It will be adopted by the OEEC, and Carlisle Humelsine testified to the Hoover Committee on July 15, 1950 concerning its provisions.

Department of State requested to provide date for notification of Civil Service Commission of reasons for resignations, March 29, 1951.

RECURSIVE SUMMARY ROUND 1

President Truman appointed James Webb as Under Secretary of State (1949-1952) and a Senate hearing revealed the Department had purged 91 homosexuals since 1947. A White House meeting was held with Webb and Senator Hoey, and Carlisle Humelsine provided material to Webb related to the security program. GATT was an agreement adopted by the OEEC and Humelsine testified to the Hoover Committee in July 1950. The Department of State requested notification of Civil Service Commission of reasons for resignations in March 1951.

RECURSIVE SUMMARY ROUND 2

President Truman appointed James Webb as Under Secretary of State (1949-1952) and a Senate hearing revealed the Department had purged 91 homosexuals since 1947. GATT was adopted and Humelsine testified to the Hoover Committee in July 1950. The Department of State requested notification of Civil Service Commission of reasons for resignations in March 1951.

RECURSIVE SUMMARY ROUND 3

President Truman appointed James Webb as Under Secretary of State (1949-1952). The Department of State had purged 91 homosexuals since 1947 and requested notification of Civil Service Commission for reasons for resignations in 1951. GATT was adopted and Humelsine testified to the Hoover Committee in 1950.

RECURSIVE SUMMARY ROUND 4

President Truman appointed James Webb as Under Secretary of State in 1949 and the Department of State purged 91 homosexuals since 1947. In 1951, GATT was adopted and Humelsine testified to the Hoover Committee.

ILLUSTRATIONS OF MOOD

Figure 1. Black-and-white pencil drawing of three men around a long boardroom table in the White House. They are all wearing suits and have stern expressions on their faces. On the table is a document detailing the State Department's purge of 91 homosexuals from its workforce. Art by herb.loc['ai'] using Stable Diffusion[1].

---

[1] The artist captures the mood of the room where decisions were taken quite nicely. –Ed.

NASA Historical Investigation into

James E. Webb's Relationship to the Lavender Scare

Final Report

submitted by

Brian C. Odom, PhD, MLIS

NASA Chief Historian

Executive Summary

The central purpose of this investigation was to locate any evidence that could indicate whether James Webb acted as a leader of or proponent for firing LGBTQ+ employees from the federal workforce.

For this purpose, the acting NASA Chief Historian examined thousands of documents at the Truman Presidential Library, as well as archival collections at NASA Headquarters, NASA's Marshall Space Flight Center, and the National Archives and Records Administration. Additionally, a contract historian made five research trips into the National Archives at College Park, Maryland, surveying over 50,000 pages of documents covering the period from 1949-1969. The report summarizes findings from these primary source documents related to James Webb's time as Under Secretary of State and NASA Administrator, as well as: secondary literature; conversations with historians and archivists who had previously studied these topics; and attempts to locate memoranda, reports, correspondence with key participants, notes, meeting minutes, or other documentation related to actions taken by Webb.

This report provides detailed context on a period in American history referred to as the "Lavender Scare"—a time characterized by the exclusion and expulsion of homosexual employees from the federal workforce starting in early 1947. While James Webb was Deputy Under Secretary of State in 1950, Congress began an investigation personnel at the State Department in the name of rooting out "the alleged employment by the departments and agencies of the Government of homosexuals and other moral perverts"—an approach solidified as executive policy under President Eisenhower in 1953.

This report closely examines two instances in which Webb enters this historical context around the Lavender Scare. One is a meeting with President Truman on June 22, 1950, to determine, in the President's words, "a proper basis for cooperation" with the Congressional investigation. The second is a meeting on June 28, 1950, with Senator Hoey, Charlie Murphy (Truman White House Counsel), and Stephen J. Spingarn (Administrative Assistant to Truman). The report details extensive primary sources around the meeting with Senator Hoey. Based upon the available evidence, Webb's main involvement was in attempting to limit Congressional access to the personnel records of the Department of State. During that meeting, Webb did pass along to Senator Hoey "some material on the subject [of homosexuality] which [Carlisle] Humelsine of State had prepared." None of the evidence found links Webb to actions emerging from this discussion. Nor does Webb, in the aftermath of the June 28 meeting, follow up on the matter – whether via memoranda or correspondence.

The report also examines Webb's time as the NASA Administrator from 1961 to 1968. By this point, the identification of the employment of homosexuals in the executive branch as a national security issue – and a fireable offense -- was executive policy under Eisenhower's 1953 Executive Order 10450 which was made policy at the federal Civil Service Commission. In 1963, Clifford J. Norton, a NASA GS-14 budget analyst, was fired due to his sexual orientation. Norton sued the Civil Service Commission. Ultimately, the 1969 federal case *Norton v. Macy* found for the appellant – one of several cases that helped pave the way for the civil service policy to be overturned, which it ultimately was in 1975.

No evidence has been located showing Webb knew of Norton's firing at the time. Because it was accepted policy across the government, the firing was, highly likely – though, sadly – considered unexceptional. We do not know if Webb knew of the *Norton v. Macy* case in 1969—there is no evidence found to support that he did.

In conclusion, to date, no available evidence directly links Webb to any actions or follow-up related to the firing of individuals for their sexual orientation. However, the research and this report make clear that the Lavender Scare was a painful chapter in our national history. Every effort was made to be as thorough in research and objective in analysis as possible. We must make great efforts to learn from the experience to guarantee that the core values of diversity, equity, accessibility, and inclusion are advanced, not only at NASA, but across the federal government. Only then can we ensure that dark episodes such as the Lavender Scare remain our history and not our future.

I. Introduction

In early March 2021, in my role as the acting NASA Chief Historian, I began an historical investigation into the career of James E. Webb at both the Department of State as the Deputy Under Secretary of State (1949-1952) and NASA as the NASA Administrator (1961-1968). In this historical investigation, I was committed to employing both a sound historical methodology and a firm commitment to objective fact and discovery. The formal effort, ordered by NASA Administrator Bill Nelson in late spring 2021, was charged with looking for evidence documenting any direct relationship between Webb and the firings of members of the LGBTQ+ community at either agency during the period now known as the Lavender Scare. Beyond a thorough investigation of the historiography on the Lavender Scare and analysis of the evidentiary record, a contract historian was brought on to attempt to locate any relevant records held at the National Archives in College Park, Maryland. NASA, Administrator Nelson, and I all recognized the importance of the issue and were sensitive to the concerns of the petitioners who brought this critical issue to our attention.

The central goal of this historical investigation was to locate any evidence indicating that during his time as Deputy Under Secretary of State, James Webb acted as an architect and leader for the firing of homosexuals from the federal workforce—an undertaking collectively known as the Lavender Scare. No attempt has been made to cover the entirety of the Lavender Scare, just those moments directly relevant to the scope of this investigation. The totality of the Lavender Scare has been covered by many other excellent historians including David Johnson (*The Lavender Scare*) and more recently by James Kirchick (*Secret City*). The main goal of this report is to utilize the available documentation to develop the historical context surrounding James Webb's time at the State Department (as relevant to the Lavender Scare), the actions of other internal managers at the State Department, and the interactions between State, the Harry Truman White House, and Congress during the years of Webb's tenure there. The second moment contextualized below is that related to the firing of Clifford Norton from NASA in 1963 and the subsequent case, *Clifford Norton v. John Macy, et. al.* (1969), a landmark ruling towards changing the Civil Service Commission's policy related to homosexuals in the federal government and curtailing the Lavender Scare.

I would like to thank the numerous historians and scholars consulted, the archivists who provided valuable insights and time to assist in locating pertinent documentation over the course of this investigation. One core takeaway from this investigation is the Lavender Scare was a dark chapter in our country's history. As with similar moments across generations, we must make great efforts to learn from the experience and work to guarantee that individual civil rights continue to be protected while positive steps are taken to ensure the goals of diversity, equity, accessibility, and inclusion are pushed forward, not only at NASA, but across the federal government. Only then can we ensure that dark episodes such as the Lavender Scare remain our history and not our future.

II. Historical Analysis

Introduction

The following is a brief outline of James Webb's career in government. In 1936 James Webb became personnel director, secretary-treasurer and later vice president of the Sperry Gyroscope Company in Brooklyn, New York, before re-entering the U.S. Marine Corps in 1944 for World War II. After World War II, Webb returned to Washington and served as executive assistant to O. Max

Gardner, then Under Secretary of the Treasury, before being named as director of the Bureau of the Budget in the Executive Office of the President, a position he held until 1949. President Harry S. Truman then asked Webb to serve as Under Secretary of State in the U.S. Department of State. When Truman left office in 1953, Webb left Washington for a position in the Kerr-McGee Oil Corp. in Oklahoma City, Oklahoma. James Webb's time in government service from 1945 to 1952 at both the United States Bureau of Budget and Department of State, as well as his time as NASA Administrator, 1961-1968, coincided with a period in American history known as the Lavender Scare—a period in which thousands of homosexual federal employees were purged from government positions due to their sexual orientation.

James Webb at State Department (1949-1952)

On February 28, 1950, deputy Under Secretary of State for Administration and Management, John Emil Peurifoy testified before the subcommittee of the Senate Committee on Appropriations that the State Department had since January 1, 1947 purged 91 homosexuals from its workforce. The revelation of these firings touched off additional Senate hearings regarding the status of other homosexuals in the federal workforce. At the time of his Congressional testimony on February 28, 1950, John Peurifoy was serving as the deputy Under Secretary of State for Administration and Management in charge of administration and security, a position which had been authorized by Congress on May 26, 1949 as part of the Department of State Organization Act of 1949 (P.L. 81-73; 63 Stat. 11).

It is important to this historical investigation to understand the context of Peurifoy's testimony. On January 21, 1950, a trial jury convicted former State Department official Alger Hiss of perjury and sentenced him to a five-year prison term. A related event was the landmark publication on January 5, 1950 of Alfred Kinsey's work, *Sexual Behavior in the Human Male*. In his work, Kinsey reported that, of his research cohort of 5,300 men, 37 percent had "reported at least one homosexual experience in their lifetime, and 10 percent were 'more or less exclusively homosexual' for a period of three years between the ages of sixteen and fifty-five."[1] Historian James Kirchick argues that Kinsey's study "inflamed" popular anxieties surrounding homosexuals. Kirchick points out that the many contemporaries began to ask the question, if the number of homosexuals were that high, how were they escaping detection?[2] Taken together, the publication of the Kinsey Report and the conviction of Alger Hiss formed a potential powder keg as the connection between communism and homosexuality began to coalesce.

The situation escalated in the aftermath of Senator Joseph McCarthy's (R-Wisconsin) February 9, 1950 address to the Women's Republican Club of Wheeling, West Virginia in which McCarthy indicated that the United States was in a "final, all-out battle between communistic atheism and Christianity." Getting more specific, McCarthy recounted, "While I cannot take the time to name all the men in the State Department who have been named as members of the Communist Party and members of a spy ring, I have here in my hand a list of 205."[3] On February 20, 1950, McCarthy, who had by now lowered his numbers from 205 to 81 communists, continued to press his claims on the floor of the United States Senate. As the day moved on, McCarthy began to blur the lines between the threats he viewed as posed by communists in the State Department to those posed by homosexuals. According to historian David Johnson, author of the *Lavender Scare,* McCarthy identified homosexuality as the "psychological maladjustment that led people toward communism. The Red Scare now had a tinge of lavender."[4]

---

[1] Kirchick, *Secret City,* 105.
[2] Ibid.
[3] Quote taken from the United States Senate webpage, https://www.senate.gov/about/powers-procedures/investigations/mccarthy-hearings/communists-in-government-service.htm.
[4] Johnson, *The Lavender Scare,* 16.

The major issue for the State Department was the perception that its internal security program was failing. These charges dated back to General George Marshall's time as Secretary of State which spanned from January 1947 to January 1949. The Senate Appropriations Committee warned Marshall in June 1947 that a "deliberate, calculated" program existed at State to protect Communist employees. In response, Marshall established a Personnel Security Board under the direct supervision of John Peurifoy which directly empowered him to remove anyone deemed a security risk. While Communists were the initial targets, the scope widened to include anyone suspected of "habitual drunkenness, sexual perversion, moral turpitude, financial irresponsibility or criminal record."[5] Under Peurifoy's watch, the State Department fired 31 homosexuals in 1947, 28 in 1948, and 31 in 1949.[6] These firings align with Peurifoy's testimony on February 28, 1950 before Congress with the majority of firings predating Webb's arrival at the State Department.

In his work *Toward Stonewall*, Nicholas Edsall argues that Lavender Scare emerged from the forces of McCarthyism "in early 1950, when an Under Secretary of state testified to a Senate committee that most of the government employees dismissed for moral turpitude were in fact homosexual."[7] Edsall was mistaken in referencing the "Under Secretary of state" as the person testifying before Congress. However, the error was not necessarily Edsall's as Senators Karl Mundt (R-North Dakota) and Kenneth Wherry (R-Nebraska) refer to John Peurifoy several times in the Congressional record as being the "Under Secretary of state" when, in reality, Peurifoy was the deputy Under Secretary of State for Administration including references listed in the Congressional

---

[5] Quotes drawn from David Johnson, *The Lavender Scare: The Cold War Persecution of Gays and Lesbians in the Federal Government.* (Chicago: Chicago University Press, 2004), 20-21.
[6] Ibid.
[7] Edsall, *Toward Stonewall,* 276.

Record on July 24, 1950, during extensive discussion of the firing of homosexual employees in the federal government.[8]

The State Department's initial response to McCarthy's charges came from deputy Under Secretary of State, John Peurifoy who issued a press release stating that "202 communists and security risks have been dismissed from the Department of State since 1946."[9] The background for this press release is also helpful in understanding the context for the upcoming testimony to the Senate on February 28, 1950. On February 28, 1950, Peurifoy and Secretary of State Dean Acheson came before the Senate Committee on Appropriations to testify "before the Subcommittee on State, Justice, Commerce Appropriations in justification of 1951 budget estimates for the Department of State in connection with the subcommittee's hearings on its titles of the omnibus appropriations bill."[10]

Another relevant exchange with bearing on this context occurred between Senator William F. Knowland (R-California), Secretary of State Dean Acheson,[11] and Peurifoy from January 26, 1950 to February 20, 1950. In a letter from Knowland, an established critic of the Truman administration, to Acheson on January 26, Knowland recalled it was the March 21, 1947 Executive Order (9835) issued by President Truman which established the procedure for loyalty investigations of government employees. In this letter, Knowland asked Acheson to determine the loyalty status of employees in the Department, particularly those with assignments in the Far Eastern Division. The foundation of this particular ask is telling. Republicans laid blame for the "loss of China" to communism the previous year squarely on the Truman Administration. In August 1949, Acheson

---

[8] Congressional Record, July 24, 1950, https://www.govinfo.gov/content/pkg/GPO-CRECB-1950-pt8/pdf/GPO-CRECB-1950-pt8-9.pdf

[9] Press release mentioned by Senator Mundt in *Congressional Record*, February 20, 1950. https://www.govinfo.gov/content/pkg/GPO-CRECB-1950-pt2/pdf/GPO-CRECB-1950-pt2-12-1.pdf

[10] *Daily Digest,* 1950 vol. 96, part 20 pages 121-125. https://www.congress.gov/bound-congressional-record/1950/02/28/daily-digest?p=0

[11] Dean Acheson served as Secretary of State from August 16, 1949 to January 20, 1953.

had issued the *China White Paper* defending the Administration, which he argued, could have done little to prevent the situation.[12] Knowing they could turn the loss of China into a political issue with voters, Senate Republicans developed a coordinated strategy to expose the "institutional failures" in the State Department and win politically in the upcoming midterm elections. On February 16, 1950, Peurifoy responded to Knowland's letter explaining the procedures at the Department for investigating new employees. Peurifoy pointed out that all employees were appointed to positions only after investigations cleared them of any association with communism. Peurifoy stated that of the 13,917 total employees, only 326 were "receiving active attention."[13] In his response on February 20, Knowland spotlighted the 326 number asking in addition to those "how many a) resigned during the course of the investigation, or b) were removed by administrative action?"[14] Knowland followed by asking:

> "When an employee resigns while under investigation or is removed as a result of the investigation, does such a notation show on his record, or is he free to go to some other government department and gain a position of responsibility without the previous facts being known to the employing agency?"[15]

Knowland highlighted the fact that someone resigning from the State Department might easily move to another federal agency with little notice. The line of argument soon elevated the issue to the Civil Service Commission. As *Lavender Scare* author David Johnson points out, Congress

---

[12] Newman, Robert P. "The Self-Inflicted Wound: The China White Paper of 1949." Prologue, *Journal of the National Archives* (14): 141–156.
[13] Congressional Record—Senate, March 4, 1950. https://www.govinfo.gov/app/details/GPO-CRECB-1950-pt2/GPO-CRECB-1950-pt2-22
[14] Ibid.
[15] Ibid.

brought pressure upon the Civil Service Commission chair, Harry B. Mitchell to investigate how many current federal employees fell under this category. Once Mitchell confirmed that many had, Congress quickly "pressed for a new policy to prevent the situation from reoccurring."[16] The Civil Service Commission responded by issuing new instructions to federal agencies requiring them to "report to the commission the specific reason for dismissals from 'suitability' charges."[17] A document located during this investigation at the National Archives in College Park, Maryland, provides a transcript of Carlisle Humelsine's July 15, 1950 testimony before the Hoey Committee.[18] Humelsine, then serving as Assistant Secretary in charge of internal security at the State Department, underneath Peurifoy, noted that the action from the Civil Service Commission to the State Department to put a procedure in place to notify that body of the specific reason for resignations was adopted at the agency on April 7, 1950. Humelsine recounted that even before this procedure was officially in place, the State Department had notified the Civil Service Commission of the reason for resignations whenever requested to do so by that body.[19]

John Peurifoy Testifies before Congress

It was the February 28, 1950 testimony of Dean Acheson and John Peurifoy before the Senate that finally brought all these disparate threads together. Historian David Johnson argues this moment had been a "political performance orchestrated by a congressional tag team intent on

---

[16] Johnson, *Lavender Scare*, 81.
[17] Ibid.
[18] Led by Senator Clyde Hoey (D-North Carolina), the Hoey Committee, spanning from July 1950 to its final report issued in December 1950, was the Senate's formal investigation of homosexuals in the federal government. This committee followed the early Senate investigation into the same subject, spanning from late March to May 1950, was led by Senators Kenneth Wherry (R-Nebraska) and J. Lister Hill (D-Alabama). The final report of the Hoey Committee concluded that homosexuals employed in the federal government did indeed constitute a security threat. This conclusion thus initiated the government's policy of firing homosexuals discovered in the workforce (formal beginning of the Lavender Scare).
[19] Testimony of Carlisle Humelsine to the Hoey Committee, July 15, 1950 pg. 2214-B
Executive Session Hearing of the Subcommittee on Investigations, Records of the U.S. Senate RG 46

assisting Senator McCarthy and embarrassing Acheson's State Department and the entire Truman administration."[20] In the aftermath of John Peurifoy's testimony, the number one concern of the State Department became minimizing the ability of the Republican led Senate to utilize any perceived failings of the security program in the State Department to score political points against the Truman White House, or worse, taking effective control of United States foreign policy by controlling the organization responsible for that policy. The available evidence positions James Webb as the person at the State Department tasked, not with expanding the scope and scale of the internal security program (that was Peurifoy and Humelsine's role), but with limiting Congress's ability to use the threat of a failure of the internal security program in the State Department to increase its insight/oversight of the State Department. The pages below provide an analysis of that evidence.

Hoey Committee – July-December 1950

Following deputy Under Secretary of State for Administration and Management John Peurifoy's testimony on February 28, 1950, an initial Senate subcommittee was established under the leadership of Senator Kenneth Wherry (R- Nebraska) and Senator J. Lister Hill (D-Alabama). This investigation lasted from March to May of 1950 and on June 7, 1950, recommended that the Senate launch a wider investigation exploring the "the alleged employment by the departments and agencies of the Government of homosexuals and other moral perverts."[21] It was the recommendations of the Wherry-Hill subcommittee that sparked the Civil Service Commission to send:

---

[20] Johnson, *Lavender Scare,* 17.
[21] Quoted in Judith Adkins, ""These People are Frightened to Death": Congressional Investigations and the Lavender Scare," *Prologue,* (Summer 2016) Vol. 48, No. 2. https://www.archives.gov/publications/prologue/2016/summer/lavender.html accessed May 16, 2021.

"Instructions to government agencies requiring them to submit detailed reasons for removals or resignations when those reasons could affect employees' suitability for reemployment so that the commission could prevent it if necessary."[22]

Civil Service Commission Chair, Harry Mitchell determined that access to the files of local police departments could assist the commission by providing lists of moral arrests that could serve as a database to assist agencies with screening current employees and future applicants—a policy that was put in place in the aftermath of the Wherry-Hill subcommittee investigation.[23] In the aftermath of the Wherry-Hill subcommittee investigation, archivist Judith Adkins argues it was at this point that the Civil Service Commission began circulating guidance to all federal departments and agencies a "letter emphasizing the necessity of reporting promptly 'the actual reasons' for all separations and resignations."[24] According to available evidence, James Webb played no role in the Wherry-Hill investigation. The Wherry-Hill investigation concluded on June 7, 1950 issuing a call for a second, more expansive investigation of "the alleged employment by the departments and agencies of the Government of homosexuals and other moral perverts."[25] The person chosen to chair this committee which began its work in July 1950 was Senator Clyde Hoey (D-North Carolina). This committee included three additional Democrats (Senators John McClellan, James Eastland, Herbert O'Connor) and three Republicans (Senators Karl Mundt, Andrew Schoeppel, and Margaret Chase Smith).[26]

[22] Lewis, "Lifting the Ban on Gays in the Civil Service: Federal Policy Toward Gay and Lesbian Employees Since the Cold War," 388-389.
[23] Ibid.
[24] Adkins, "These People are Frightened to Death," *Prologue Magazine*, Vol. 48, No. 2 (Summer 2016).
[25] Ibid.
[26] Ibid.

The spring of 1950 was a critical moment for United States foreign policy in the face of a rapidly escalating Cold War. On April 7, 1950, the Department of State's Policy Planning Staff completed National Security Council Paper NSC-68 entitled "United States Objectives and Programs for National Security." This Top-Secret report, potentially the most influential United States document drafted during the Cold War, recommended a massive buildup of conventional and nuclear arms as the only way to deter the Soviet threat.[27] Giving credence to this strategy was the outbreak of the Korean War on June 25, 1950. The United States entered the war two days later on June 27, 1950.

In terms of the political context, it is important to remember that the charges of lax security were leveled against the State Department (an agency of the Truman Administration) by Republican members of the Senate. Limiting Congressional oversight of and insight into the workings of a government agency appears to be James Webb's primary motivation for getting involved with the Hoey Committee. Based upon the available evidence, Webb enters the historical context surrounding the Lavender Scare at two primary moments in the historical record: 1) a June 22, 1950, meeting with President Truman to determine, in the President's words, "a proper basis for cooperation" with the Congressional investigation, and 2) a June 28, 1950 meeting with Senator Hoey, James Webb, Charlie Murphy (Truman White House Counsel), and Stephen J. Spingarn (Administrative Assistant to Truman).

These two moments merit considerable attention as they are central to this investigation. In the first moment—Webb's June 22 meeting with Truman—the central questions are why was Webb meeting with Truman in this instance? Why was coordination with the White House important to what would follow? What were the central points under consideration? What strategy was agreed

---

[27] "NSC-68, 1950" Office of the Historian, Department of State. https://history.state.gov/milestones/1945-1952/NSC68

upon for working with the Hoey Committee? In the second moment—Webb's June 28 meeting

with Senator Hoey, White House Counsel Murphy, Truman's Administrative Assistant Spingarn—

the key questions are what was the basis of the material Webb handed to Senator Hoey? What was

Webb's objective at the meeting? What strategy was agreed upon with Senator Hoey? Taken

together, these two moments are key in understanding Webb's actions:

• Webb passed along Carlisle Humelsine's (head of internal security at the State

Department, then Deputy Under Secretary of State for Administration and Management once John

Peurifoy left the Department) memorandum on homosexuality to Senator Hoey during their June

28, 1950 meeting, a report Stephen J. Spingarn who was then serving as Administrative Assistant to

Harry Truman at the White House. Correspondence between the State Department and the Truman

White House indicates that the primary issues discussed concerned limiting access to files and names

of agency personnel and the nature of the Committee's hearings (public or executive session).

• Based upon the available evidence, my analysis will show that James Webb's primary

concern in the matter was to limit Congressional involvement/access to the personnel records of the

Department of State—something he was asked by President Truman to ensure.[28]

On June 22, 1950, with Secretary Dean Acheson out of town, James Webb attended the

standing Thursday meeting with President Truman. David Johnson recounts that during their

meeting, Truman and Webb discussed a strategy of engagement with the Hoey Committee

determining how they might "work together on the homosexual investigation" with Truman

commenting that "he was sure we could find a proper basis for cooperation."[29] Truman directed

---

[28] Documentation cited in David Johnson, *The Lavender Scare* as: David D. Lloyd to Mr. Spingarn, July 3, 1950, and Stephen Spingarn, "Memorandum for the Hoey Subcommittee Sex Pervert Investigation File," June 29, 1950, both in "Sex Perversion" Folder, Box, 32 WHCF, HST Library; James E. Webb, "Meeting with the President, Thursday, June 22, 1950," Box 9, Entry 53D444, Secretary's Memoranda, 1949-1951, Records of the Executive Secretariat, RG 59, NARA.
[29] Johnson, *Lavender Scare,* 104.

Webb and two White House aids, Stephen Spingarn and Charlie Murphy, to meet with Hoey.[30]

Johnson notes that over the following weeks, it was Spingarn and Murphy who "met repeatedly"

with Hoey Committee Chief Counsel, Francis Flanagan. It was Flanagan who led the real work of

the Committee including research, witness selection, and drafting the final report.

Between James Webb's June 22 meeting with Truman and his June 28 meeting with Senator

Hoey, State Department Security Officer Carlisle Humelsine (who reported to John Peurifoy)

gathered a packet of material related to the security program—providing it to Webb on June 24,

1950. The packet consisted of five memoranda including: a background paper on the "problem of

homosexuals and sex perverts in the Department of State"; Arch Jean's (Chief of State Department

Personnel) report to his supervisor Peurifoy of a meeting with Francis Flanagan on June 20, 1950;

suggestions "as to the objectives of the [Hoey] Committee and Methods of Operation; a

memorandum "suggesting the organization and principles to govern the Department's participation

in the Senate inquiry;" and finally, a list of the Hoey Senate Committee members.[31]

The most instructive memorandum included in Humelsine's packet to Webb was the

"Report of Meeting with Mr. Flannagan [sic], Senate Investigations Staff." This memorandum

recounts Francis Flanagan's (Chief Counsel for the Hoey Committee) instructions to the State

Department as to what information would be needed and highlights the Hoey Committee's desire to

conduct its charge with little public attention or politization. Drafted by Arch Jean, Chief of

Departmental Personnel to John Peurifoy on June 20, 1950 following a visit with Flanagan, the

memorandum provides a roadmap to both the requests made upon the State Department from the

Committee as well as its stance on key issues. Jean reported that his meeting with Flanagan was

positive as Jean found him to be a "personable individual" searching for "ways and means to

---

[30] Ibid.
[31] Memorandum for Mr. Webb, June 24, 1950.

accomplish his task without fanfare and without embarrassment to the agencies or the people involved."[32] Jean then listed the main points from his meeting with Flanagan that would need to be considered by the State Department in its future dealings with the Hoey Committee. These points included: a) Flanagan's request for disclosure of related State Department files to the Hoey Committee to use "as they deem proper and necessary"; b) need for a statement including State Department views on homosexuality from a "sociological standpoint"; c) list of State Department personnel who were "well informed on the subject" and could be called to testify before the Committee; d) statistical data, dating back to July 1, 1945, on how many employees had resigned, been dismissed, and were currently under investigation; e) a detailed overview of the security procedure in place at the State Department; and f) examples of "homosexuals' tendency to locate employment with others of their kind in the same agencies. Jean passed along Flanagan's desire to conduct "most, if not all, of the hearings in executive session."[33] From his meeting with Flanagan, Jean came away with the strong impression that Flanagan had "already concluded that homosexuals should not be employed in government under any circumstances."[34]

The memorandum "Problem of Homosexuals and Sex Perverts in the Department of State," drafted by Carlisle Humelsine, written in response to Flanagan's meeting with Arch Jean from June 20, 1950 due to Humelsine's narrative providing a nearly point by point reply to the Committee's request. In this memorandum, Humelsine provided the Department's views on homosexuality from Flanagan's requested 'sociological standpoint.' Humelsine then recounts the history of the State Department's evolving stance toward homosexuals over the years saying that "until very recent years" the department and "several agencies of the Federal Government" had "tolerated homosexuals in its employment solely because not much was known about them or who they

---

[32] Arch Jean to John Peurifoy, "Report of Meeting with Mr. Flannagan, Senate Investigation Staff, June 20, 1950.
[33] Ibid.
[34] Ibid.

were."[35] For Humelsine, a lack of engagement with the subject had allowed numbers of homosexuals in the Department to expand. Humelsine points to January 1947 (the date Jean mentioned in his memo) and Peurifoy's rise to "Assistant Secretary for Administration" as the moment "homosexuality in the Department of State was dealt with in a direct and forthright manner."[36] This is consistent with previous testimony from Peurifoy on the subject and aligns with policy emanating from both the executive branch and Congress at that time. Humelsine then pointed to "Civil Service rules" that prevented the "appointment of anyone who is guilty of 'criminal, infamous, dishonest, immoral, or notoriously disgraceful conduct.'" (emphasis in original)[37]

Following a narrative of the characteristics he believed made homosexuals "undesirable as employees," Humelsine provided a list of reasons why they might seek employment. Humelsine explored the workings of his Security Division and provided an overview of the investigatory process by which claims were examined. Within the Security Division were two full time investigators charged with both detecting homosexuals and "study[ing] the problem."[38] These investigations included extensive background checks, numerous interviews with anyone familiar with the individual, surveillance, and a personal interview with both the investigator and "often by the Chief of either the Division of Departmental Personnel or Foreign Service Personnel, depending upon the service in which he is employed." Individuals determined to be homosexual either by "investigation or admission" were "promptly separated from the Department."[39] While Humelsine considered homosexuals to be "weak, unstable and fickle people who fear detection and who are therefore susceptible to the wanton designs of others," he argued there was no evidence to the fact

---

[35] Humelsine to Webb, "Problem of Homosexuals and Sex Perverts in the Department of State," undated.
[36] Ibid.
[37] Ibid.
[38] Ibid.
[39] Ibid.

that "these designs of others have caused a breach of the security of the Department."[40] Humelsine also understood that the nature of such claims against individuals created opportunities of "possible malicious charges."[41] Within the historical context of McCarthy's charges against the department, Humelsine's memorandum served several ends. The memorandum demonstrated a) the State Department security program's philosophical alignment with Congressional statements on the 'dangers' of homosexuals in the federal workforce and b) the mechanisms in place for both discovery and termination. These two points would have been the primary value of the memorandum Humelsine provided to Webb for his meeting with Hoey.

Two additional memoranda from Humelsine's packet to Webb are also instructive of the State Department *modus operandi* for working with the Hoey Committee. The first, "Suggestions as to the objectives of the Committee and Methods of operation," breaks down the recommended operation of the Hoey Committee by listing out overall objectives and suggested procedure. On objectives, the State Department recommended the Committee "evaluate the problem in its totality" and avoid trying to "determine the innocence or guilt of individual employees." They also recommended a focus upon a "study of present conditions" as related to "policies and procedures in the several agencies" while avoiding "digging up individual cases which have been handled in the past."[42] In both instances, the State Department wished to avoid a *witch trial* affair or any investigations opening State Department workforce to individual claims by conducting the hearings in the *abstract*. The memorandum listed principle questions the Hoey Committee should consider in its task of determining what "specific administrative, security and other problems, present or potential" were "posed by homosexuals in the federal government," namely: a) were homosexuals

---

[40] Ibid.
[41] Ibid.
[42] "Suggestions as to the objectives of the [Hoey] Committee and Methods of operation," Confidential Files (Truman Administration), 1938-1953) Sex Perversion [Investigation of Federal employees]
https://catalog.archives.gov/id/54538193

security risks? b) if they were determined to be a security risk at a "sensitive agency" such as the State Department, "should they be employed in a non-sensitive agency?" c) if there were not a security risk, should they be "employed in the federal government as a matter of policy?" and d) what course was "dictated by the best medical judgement?"[43]

Under suggestions for committee procedure, Humelsine recommended several points—primarily that the Committee proceedings should be "free of partisan thinking and action," held in executive session, and that official liaisons should be designated from each Department and Agency. Possibly the most important recommendations came in point five in which Humelsine suggested the Committee refrain from "reviewing individual cases." Clearly, the White House wanted to avoid allowing Congressional Republicans any chance to turn the proceedings into a public forum as they had over claims of communists in the Departments and Agencies. Humelsine also reiterated the White House's opposition to providing personnel records or investigation files to the Committee. Moving forward, this would be, as it had been in the past, the primary point of contention between Congressional Republicans and the White House.[44]

The second of these additional memoranda, the "Organization and principles to govern the Department's participation in the Committee inquiry," sought to establish the working relationship to the Hoey Committee and State Department. The memorandum listed four steps to be followed. First, the Deputy Under Secretary for Administration (Peurifoy) or his Deputy (Humelsine) would serve as the "Department's spokesman" to the Hoey Committee, members of Congress, and the press.[45] This point gave Peurifoy, then Humelsine, blanket authority for "all actions and

---

[43] Ibid.

[44] Ibid.

[45] It is important to note that Peurifoy's appointment as Deputy Under Secretary of State for Administration and Management ended on August 10, 1950 when he was named Ambassador Extraordinary and Plenipotentiary to Greece. He was replaced on that date by his deputy, Humelsine.

pronouncements of the Department relating to this subject."[46] Acheson and Webb would be "kept

informed of all significant developments" and would be "available for behind the scene activities."

There is no indication in the available evidence that either Acheson or Webb were brought back into

the activities in any way beyond occasional updates at staff meetings.[47] Secondly, the memorandum

called for an "Ad Hoc Committee" to be created for Peurifoy and Humelsine to "serve as a

sounding board and to advise him [Peurifoy] on courses of action under varying circumstances."[48]

The creation of this ad hoc committee suggests that advice and counsel would take place outside the

purview of Acheson and Webb. A third point suggested Peurifoy and Humelsine provide the

Committee with "such statistics as will be useful" without "jeopardizing the Department's personnel

and security programs." This point suggests that the primary concern remained limiting the scope

(insight) of the Hoey Committee into the affairs of the State Department rather that identifying and

terminating homosexual employees. Finally, a fourth point reiterated this objective stating that the

State Department would "resist any attempt of the [Hoey] Committee to obtain names of

individuals and files."[49] It is important to note that the number one concern was not with identifying

homosexual employees in the workforce but instead limiting the overall scope of the Hoey

Committee (Congress) to gain insight into the security program at the State Department. At no

point does Webb or Acheson express a willingness to expand the scope of the security program

within the Department.[50]

Subsequently on June 28, 1950, Webb participated in a meeting with Charlie Murphy, Steven

J. Spingarn, and North Carolina Senator Clyde Hoey. During that meeting, Webb passed along to

---

[46] "Memo suggesting organization and principles for the Department's participation in the Senate Inquiry," Confidential Files (Truman Administration), 1938-1953) Sex Perversion [Investigation of Federal employees] https://catalog.archives.gov/id/54538193
[47] Ibid.
[48] Ibid.
[49] Ibid.
[50] Ibid.

Senator Hoey "some material on the subject [of homosexuality] which [Carlisle] Humelsine of State had prepared." To date, no available evidence directly links Webb to any actions emerging from this discussion, notably any actions later undertaken by either the Department of State or the Hoey Committee. The State Department determined that the dealings with the Hoey Committee would be led by the deputy Under Secretary of State for Administration and Management (Peurifoy, then Humelsine) and that the Secretary and Under Secretary would only be made available as needed. Because of this, it is a sound conjecture that Webb played little role in the matter, from either an administrative or philosophical perspective, beyond the June 28, 1950 meeting with Senator Hoey. The absence of any reports, correspondence, memoranda, etc. in the historical record backs that assumption.

The key question then remains: what was Webb's primary reason for attending this meeting with Hoey, Spingarn, and Murphy? An analysis of the available evidence, historical context, and resulting actions from the meetings underscores the point that Webb's goal was acting on behalf of President Truman to draw a line between Congress and the State Department. President Truman's main concern with McCarthy's claims against the State Department was that it would result in increased Congressional insight/oversight of foreign policy at such a critical moment in international affairs. By claiming there were problems with the security program in the State Department, McCarthy placed the Administration in a delicate position. Truman's response to the current attempt from the Senate reflected his response to prior attempts from the House Committee on Un-American Activities (HUAC) in March 1948 to gain access to personnel records. That instance involved a HUAC subpoena to the Secretary of Commerce to produce files related to the loyalty investigation of Dr. Edward Condon. Truman responded on March 13, 1948 with a directive protecting all such files from subpoenas or demands in accordance with Executive Order 9835 (March 21, 1947). In the language of the directive, all such requests for files "shall be respectfully declined" with any such requests being "referred to the Office of the President for such response as the President may determine to be in the public interest in the particular case."[51] At an April 22, 1948 press conference, Truman reiterated his refusal to turn over such papers to HUAC.

A year later, Truman again used this precedent in his refusal to turn over files to the subcommittee of the Senate Foreign Relations Committee investigating the claims of disloyalty in the State Department. The March 28, 1950 subpoena from the Committee was denied by President Truman on April 3, 1950 pursuant to the March 13, 1950 directive. Due to the public nature of McCarthy's claims toward State Department personnel and with the upcoming Hoey Committee investigation into homosexual employees in the federal government, Truman with a political problem. Turman fell back on the March 13, 1948 directive refusing to open State Department security investigation files to Congress.[52]

The meeting by Senator Hoey with Webb, Spingarn, and Murphy on June 28, 1950 concerned ensuring compliance with that precedent. This is evident in the correspondence unfolding in the aftermath of the meeting. Once Senator Hoey, and later Flanagan, agreed to proceed under this understanding, discussions with the Hoey Committee or any documentation surrounding the security program no longer include Webb directly. The evidentiary record supports the claim that the primary concern of the White House, and subsequently the Departments and Agencies, was not with expanding the scope of the homosexual investigations but diffusing the ability of the Congressional Republicans to use the issue as they had done with potential communists in the executive branch. In memoranda and correspondence from White House staff, including Charlie Murphy, Donald Dawson, and Stephen Spingarn, the primary objective of the White house is to

---

[51] Federal Register, Volume 13, Number 52, March 16, 1948. https://tile.loc.gov/storage-services/service/ll/fedreg/fr013/fr013052/fr013052.pdf

[52] Theodore B. Olson, "History of Refusals by Executive Branch Officials to Provide Information Demanded by Congress, December 14, 1982. https://www.justice.gov/file/23246/download

limit the scope of the Committee by ensuring personnel files were withheld from the proceedings.

Examples of this are found in Stephen Spingarn's June 29, 1950 memoranda to Donald Dawson

concerned with "disregarding these requests from Flanagan"[53] and another from Spingarn to

Dawson on the same day providing a fuller account of the meeting with Webb, Murphy, and Hoey.[54]

This second memorandum reiterates the executive branch's primary concern of limiting the

scope of the Committee's insight into the security program of the departments and agencies. Here,

Spingarn pointed out the meeting's primary agenda aims of limiting testimony to the security

program to the abstract (testimony of medical authorities and senior departmental security officers)

and preventing access to departmental names or files. Spingarn recounted that during the meeting

with Hoey, Charlie Murphy expressed this desire noting that if such a call did come from the

Committee, it would be denied "on the basis of the 1948 Presidential directives." On the issue of

holding some part of the hearings in public, notably the medical testimony, Spingarn observed that

Murphy was for holding all in private (executive) while Webb was "not certain." Webb's opinion

here implies that he had not given that aspect much thought before the meeting—a fact that

supports the argument that Webb's purpose for being at the meeting was simply to ensure the

departmental files were closed to Congress, something he and President Truman would have made a

primary concern on the scope of the committee.[55]

Two additional memoranda located in the Stephen J. Spingarn Papers at the Truman

Presidential Library add additional context for the aftermath of the June 28 meeting with Hoey. The

first, written on July 5, 1950 by Spingarn, references a follow-up meeting between Spingarn and

---

[53] Memorandum from Stephen J. Spingarn to Donald S. Dawson with Note, Office of the President, 4/1945-1/20/1953, File Unit: Sex Perversion [investigations of Federal employees], 1945 – 1953, Confidential Subject Files, 1945 – 1953.

[54] Note from Stephen J. Spingarn to Donald S. Dawson with Attached Memorandum, Office of the President. 4/1945-1/20/1953, Sex Perversion [investigations of Federal employees], 1945 – 1953, Confidential Files (Truman Administration), 1938 – 1953.

[55] Ibid.

Hoey on July 5 at the Capitol. There, Spingarn clarified the earlier question of whether any aspect of the Committee's work should be conducted in public. Spingarn indicated that it was the "unanimous opinion of the White House staff…that all of the hearings, including the medical testimony, should be in executive session."[56] Hoey informed Spingarn that he hoped to follow that plan but thought there "might be some opposition from the Republicans, particularly Senator Mundt."[57] Hoey also brought up the point that the Committee would be collecting the arrest records of the Washington Police which they would then contact the departments for additional information. Spingarn stated this would "present some problems on the disclosure of information" but that he did not "see how it can be avoided since the police records are public records and are not within the President's directives about non-disclosure of personnel files and information."[58]

A subsequent memorandum from July 10, 1950 details a meeting between Spingarn and Francis Flanagan (Chief Counsel of the Hoey Subcommittee) during which Spingarn reiterated the White House's desire for non-disclosure of personnel files. At the request of Hoey, Flanagan met with Spingarn to discuss the operations of the committee. Spingarn recounted that he "went over the same ground with him that Mr. Webb, Mr. Murphy, and I had gone over with Mr. Hoey."[59] This is consistent with David Johnson's argument that it was Hoey's discomfort with the topic and hope not to "have any hearings that McCarthy can make big headlines out of" that made Flanagan the "driving force behind the Hoey Committee's investigation."[60] In the July 10 memorandum, Spingarn's notes support this claim stating that although Flanagan was not able to "produce much

---

[56] Stephen J. Spingarn, "Memorandum for the Hoey Subcommittee Sex Pervert Investigation File, July 5, 1950. Stephen J. Spingarn Papers Box 13, Assistant to the President File, Chronological File, July – August 1950, Truman Presidential Library.
[57] Ibid.
[58] Ibid.
[59] Stephen J. Spingarn, "Memorandum for the Hoey Subcommittee Sex Pervert Investigation File, July 10, 1950. Stephen J. Spingarn Papers Box 13, Assistant to the President File, Chronological File, July – August 1950 Folder, Truman Presidential Library.
[60] Johnson, *Lavender Scare,* 102-103.

dope in documented instances in which homosexuals had endangered security," that he remained "convinced that homosexuals represented a serious security threat."[61] Spingarn alluded that during this conversation, he suggested that the security threat posed by homosexuals should be "squared up against other types of security threats by individuals resulting from normal sexual or non-sexual activity." Spingarn stated that that argument "did not seem to impress him much."[62]

Flanagan again raised the issue of access to names and files from the departments and agencies which Spingarn brushed aside by saying they might provide "memoranda summarizing individual files without mentioning any names."[63] In both cases, Flanagan argued that might satisfy him but not Senator Mundt and other Republican members of the Committee. Flanagan also indicated that Mundt would see in this refusal of access or sampling of data the agencies not giving an "accurate account of what was in the files," something Spingarn noted was "what Senator McCarthy said about the loyalty files."[64] Spingarn also recounted his impression that the White House might have "some difficulties" from Flanagan due to the fact that he "seems to have pre-judged one of the central issues that the Subcommittee has to decide, namely, how serious the threat is of the homosexual employees, particularly in relationship to other types of security threats." In Spingarn's opinion, Flanagan seemed "strongly committed to the position that the homosexual is the most serious security threat of all" and regarded the lack of evidence of it as "an unfortunate accident." Spingarn recalled that in his own experience working in counter espionage and as a security officer during World War II, he was "personally of the opinion (I believe I can cite chapter and verse to support it) that other types of security threats are more dangerous than homosexuals,

---

[61] Stephen J. Spingarn, "Memorandum for the Hoey Subcommittee Sex Pervert Investigation File, July 10, 1950. Stephen J. Spingarn Papers Box 13, Assistant to the President File, Chronological File, July – August 1950 Folder, Truman Presidential Library.
[62] Ibid.
[63] Ibid.
[64] Ibid.

although no doubt he represents one."[65] Spingarn's discussion with Flanagan closely resembled the experience of Arch Jean (detailed above) on June 20, 1950. Jean also observed that while he was convinced of Flanagan's "sincerity to conduct an intelligent, non-political investigation," he was also sure that Flanagan had "already concluded that homosexuals should not be employed in government under any circumstances."[66]

This is an interesting position and one I think stands at the core of the Truman White House's engagement with the Hoey Committee. Because Flanagan viewed homosexuals as major security risk, he was able to use the threat of Congressional Republicans to achieve this end with the Hoey Committee. If Spingarn, the person tasked by the Truman Administration to work directly with the Hoey Committee, held this opinion on the security threat posed by homosexuals in the federal workforce, it seems highly unlikely he would have been selected for that position if his opinion on the matter was antithetical to Truman's own. Not that anyone in the Administration showed any interest in defending homosexuals however limiting the scope of the proceedings and access to department and agency personnel files was much more the central issue for Truman, Spingarn, Murphy, and, in his own involvement in the matter, Webb's.

Following the June 28 meeting, no memoranda or correspondence has been located in which Webb follows up on the matter in any way. It is logical to expect that he would have been briefed on the topic as needed or requested, but as the Hoey Committee eventually agreed to proceed along the lines suggested by the White House, there would have been little reason for Webb to return to the issue.

One important document located in the James Webb Papers at the Truman Presidential Library reveals James Webb's prioritization on the day of his meeting with Hoey. Planning to leave

---

[65] Ibid.
[66] Arch Jean to John Peurifoy, "Report of Meeting with Mr. Flannagan, Senate Investigation Staff, June 20, 1950.

for vacation to North Carolina the following morning (June 29), Webb left a memorandum to Secretary Acheson discussing the most critical aspects of that week's business related to the "organization and administration of the Department of State."[67] In the memorandum, Webb highlighted the priorities for that week noting needed assignment of work responsibilities for William Harriman, Dean Rusk, and Charles Bohlen all of whom had just returned to the country. Webb went into detail noting the suggested assignments for each official. At no point does Webb mention his meeting with Hoey, discuss any security issues, or point to next steps in relations with Congress or the Hoey Committee. Webb left government service in 1952 returning to work in the private sector.

Administrative Changes at the State Department post-Webb

Relevant to this investigation was Congressional testimony from then Under Secretary of State for Administration Carlisle Humelsine before the Committee on Foreign Affairs in the House of Representatives on January 28, 1953. The occasion of the testimony was an hearing informing the decision to amend section 1 of the Act of May 26, 1949 adding a second Under Secretary of State (for Administration) to be appointed by the President and confirmed by the Senate. During that testimony, Humelsine pointed out the traditional roles and responsibilities of his position in relation to that of both the Secretary of State and the Under Secretary of State. Humelsine reveals that traditionally the Secretary of State was out of the country 50-60% of the time, and that during that time, the Under Secretary of State served as the acting Secretary of State. Humelsine pointed out that under these circumstances, "the regular Under Secretary, the single Under Secretary that we

---

[67] James E. Webb to Secretary Dean Acheson, June 28, 1950, James E. Webb Papers Box 24, Notes on Conversation with Secretary of State Organization and Administration of the Department of State Folder, Truman Presidential Library.

now have, is the Acting Secretary of State at least half the time" and that that was true when Webb was in the position. For Humelsine, this meant that there had been "no operating Under Secretary over 50 percent of the time."[68] Humelsine's claim is that up to this point, he and before him, John Peurifoy were leading the internal administration of the Department of State from their positions as assistant Under Secretary of State for Administration. This change to previous appropriations would rectify the imbalance by making the position consummate to its responsibilities and placing the position above the other assistant secretaries in the department.

This is important to this investigation in that it repositions responsibility for internal loyalty and security programs with that position and not the Under Secretary of State. Humelsine hints at as much in an answer to Franklin D. Roosevelt Jr. (D-NY). There, Humelsine claims that the primary job of the Under Secretary of State (position held by Webb) was to maintain a familiarity with foreign policy "getting himself so acquainted with policy that when the Secretary is away, he can take over that policy responsibility." It would now be the job of the assistant Under Secretary of State (positions held by Peurifoy, then Humelsine), as Humelsine recounted it had been in the past, to spend "his entire time trying to organize, reorganize," ensure "that the Department functions correctly," and to "look into such things as the loyalty program of the Department and make sure that there is a proper program to assure there is a loyal group of employees."[69] Clearly, Humelsine was noting that this had always been the mode of operation in the Department and that in strengthening earlier staffing appropriations from Congress, that position could devote 100 percent of its time to such matters. As Humelsine continued, during his occupancy of the position, he had

---

[68] Providing for an Under Secretary of State for Administration. Hearing before the Committee on Foreign Affairs, House of Representatives, Eighty-Third Congress First Session on S. 243 and H.R. 1377, Bills to Provide for an Under Secretary of State for Administration, January 28, 1953.
https://www.google.com/books/edition/Providing_for_an_Under
Secretary_of_State/FnmgieGkJQ8C?hl=en&gbpv=1&dq=providing+for+an+Under
Secretary+of+state&printsec=frontcover
[69] Ibid.

not "found enough hours in a day to spend much time on the question of reorganizing the Department of State."[70] Humelsine also pointed out that the potential appropriations were timely as it appeared the "State Department is going to be the most carefully investigated department in the history of the United States" due to investigations from both the House of Representatives and the Senate—the latter of which he mentioned was, with the Foreign Relations Committee of the Senate, "creating a loyalty subcommittee that is going into the loyalty and security program of the Department of State."[71] Humelsine recognized that the position he currently occupied and was working to find appropriations for, would be responsive to this continued scrutiny from the Senate.

With the election of Dwight Eisenhower to the Presidency in 1952, the identification of the employment of homosexuals in the executive branch as a national security issue was solidified as executive policy with Eisenhower's Executive Order 10450. Here, Eisenhower stated:

> "WHEREAS the interests of the national security require that all persons privileged to be employed in the departments and agencies of the Government, shall be reliable, trustworthy, of good conduct and character, and of complete and unswerving loyalty to the United States;"[72]

The order made explicit what was previously implicit in adding the category "sexual perversion" to the information to be considered in security investigations. David Johnson argues that while previous usage in civil service policy of the language "criminal" and "immoral" had "already been used to bar homosexuals—the inclusion of the more specific reference to 'sexual perversion' was

---

[70] Ibid.
[71] Ibid.
[72] Executive Order 10450 – Security Requirements for Government Employment. April 27, 1953. https://www.archives.gov/federal-register/codification/executive-order/10450.html

unprecedented."[73] Executive Order 10450, with the inclusion of 'sexual perversion' in the security criterion, remained federal policy until the aftermath of the decisions in *Bruce Scott v. John Macy, et al.* (1965) and *Clifford Norton v. John Macy, et. al.* (1969).

Webb at NASA— Clifford Norton v. John Macy, et. al. (1969)

James Webb returned to Washington on February 14, 1961, when he accepted the position of administrator of NASA. Under his direction the agency undertook the goal of landing an American on the Moon before the end of the decade through the execution of Project Apollo. For seven years from February 1961 to October 1968, James Webb served as the NASA Administrator. During his time, Webb worked diligently to enforce President Kennedy, and later President Johnson's, goals surrounding equal employment opportunity and civil rights at the agency. However, his time as NASA Administrator also occurred during the events leading to the United States Court of Appeals for the District of Columbia decision in *Clifford Norton v. John Macy, et al.* (1969).

This case originated from the firing of Clifford Norton in 1963 due to his arrest for a homosexual act. Norton, a GS-14 budget analyst at NASA, was arrested for a minor traffic violation in the early morning hours of October 22, 1963 near Lafayette Square in Washington, D.C. That morning, Norton and Madison Monroe Proctor were observed by two DC Moral Squad officers who followed both men, then driving separate cars, to Norton's residence Southwest Washington apartment parking lot. There, Proctor told the two officers that Norton had "felt his leg" and extended an invitation to his apartment.[74] Both men were arrested and taken to the DC Morals Office for further questioning. Following two-hours of questioning, the head of the Morals Squad,

---

[73] Johnson, *Lavender Scare,* 123.
[74] Clifford L. Norton, Appellant, v. John Macy, et al., Appellees, 417 F.2d 1161 (D.C. Cir. 1969). https://law.justia.com/cases/federal/appellate-courts/F2/417/1161/190082/#fn29_ref

Roy Blick, called NASA Security Chief, Bart Fugler, who arrived at the DC Morals Squad Office at 3:00am. Fugler was allowed to read the arrest record and watch a twenty-minute interrogation of Norton.[75]

During this time, Norton denied any homosexual advances to Proctor during the encounter. Norton was given a traffic ticket by the Morals Squad and released. Norton was then taken by Fugler to the "Tempo L" building where he was questioned by Fugler until 6:00am. During this questioning, Norton recalled that he had experienced certain homosexual activities in high school and college and that:

> "He sometimes experienced homosexual desires while drinking, that on rare occasions he had undergone a temporary blackout after drinking, and that on two such occasions he suspected he might have engaged in some sort of homosexual activity."[76]

Norton recounted that he had experienced a similar blackout that evening when he met Procter, although he recalled "only that he invited the man up for a drink."[77] Fugler (NASA Security Chief) determined from his investigation that Norton's actions "amounted to 'immoral, indecent, and disgraceful conduct'"—a fireable offense under the guidelines of the Civil Service Commission.[78] Norton appealed his firing.[79] The decision was reviewed, not by Webb, but by a Civil Service Appeals Examiner and the Board of Appeals and Review both of which upheld the firing.[80]

---

[75] Ibid.
[76] Ibid.
[77] Ibid.
[78] Ibid.
[79] Ibid.
[80] Ibid.

Clifford Norton sought redress of this action through the federal court system resulting in the 1969 ruling in *Clifford Norton v. John Macy Jr. et. al.* The Court's 1969 ruling in favor of Norton's claim—a landmark decision in terms of homosexual rights in federal service. In his majority opinion, Chief Circuit Judge David L. Bazelon underscored the point that, because Norton was veterans' preference eligible, he could only be fired for "such cause as will promote the efficiency of the service."[81] Previous rulings in federal courts concluded the Civil Service Commission did enjoy "wide discretion in determining what reasons may justify removal of a federal employee."[82] However, Judge Bazelon argued that "since the record before us does not suggest any reasonable connection between the evidence against him and the efficiency of the service," the court could conclude that Norton was "unlawfully discharged."[83] Numerous research efforts of NASA History archival collections, those at the National Archives, and related repositories have turned over no direct evidence that Webb ever knew anything about Norton's firing from the agency as the action taken against Norton was consistent with civil service policy. The action against Norton was, as mentioned by his boss Robert F. Garbarini, "custom within the agency" at the time he was fired.[84] Garbarini came to this conclusion after talking with "advisors" in the role of NASA personnel officers. This is certainly true given that was federal policy at the time in alignment with Executive Order 10450 (1953) and procedures put in place by the Civil Service Commission.

By the time the court ruled in Norton's favor, instituting a change to government policy, Webb had moved on from the agency, resigning from NASA on October 7, 1968 in the aftermath of the investigation of the tragic Apollo 1 tragedy. As Norton's firing from the agency was in line

---

[81] Clifford L. Norton, Appellant, v. John Macy, et al., Appellees, 417 F.2d 1161 (D.C. Cir. 1969). https://law.justia.com/cases/federal/appellate-courts/F2/417/1161/190082/#fn29_ref
[82] Ibid.
[83] Ibid.
[84] Ibid. Robert F. Garbarini was the NASA Director of Engineering in the Office of Space Science Applications in 1964 and the NASA Deputy Associate Administrator for Space Science and Applications in 1966.

with Civil Service Commission policy and federal guidance (Executive Order 10450), it is unlikely that the issue was ever presented to Webb in his role as NASA Administrator.

It is worth noting that Norton's suit was not against NASA, but John W. Macy Jr., executive director of the Civil Service Commission from 1953 to 1958 and chairman of the Civil Service Commission from 1961 to 1969. Pioneering gay-rights activist and astronomer Frank Kameny made the Civil Service Commission and its policy the focus of his and his allies fight protect the rights of LGBTQ+ regarding federal jobs. In a February 25, 1966 letter to the Mattachine Society (a national gay rights organization), John Macy spelled out the Commission's policy noting:

> "Persons about whom there is evidence that they have engaged in or solicited others to engage in homosexual or sexually perverted acts with them, without evidence of rehabilitation, are not suitable for Federal employment."[85]

Historians generally couple the decision in Norton with the same court's earlier decision in *Scott v. Macy (1965)*. In *Scott*, the policy dictating Bruce Scott's disqualification "for employment in the competitive service because of immoral conduct" was spelled out in Civil Service Regulations, 5 C.F.R. § 2.106 (1961 ed.):

> " Disqualifications of applicants.
>
> (a) Grounds for disqualification. An applicant may be denied examination and an eligible may be denied appointment for any of the following reasons:

---

[85] Quoted in "Federal Employment of Homosexuals: Narrowing the Efficiency Standard," *Catholic University Law Review*, Vol. 19:2 (1970).

...(3) Criminal, infamous, dishonest, immoral, or notoriously disgraceful conduct;"[86]

In his opinion in *Scott*, Chief Judge Bazelon contended that with its conclusion that Scott had engaged in "immoral conduct," the Commission had "not only disqualified him from the vast field of all employment dominated by the Government, but also jeopardized his ability to find employment elsewhere."[87] The ruling in *Scott* did not overturn the policy of denying homosexuals employment in the civil service, it only demanded the Commission "define its terms and 'at least specify the conduct it finds 'immoral'" while placing a greater burden of evidence on its claims.[88] Paired with the later verdict in *Norton,* which called upon the Commission to demonstrate how the excluding or firing of homosexuals from the civil service could "promote the efficiency of the service," these two cases provided a foundation for a reversal of the Civil Service Commission's policy. As argued by historian David Johnson, it was the 1973 decision in *Society for Individual Rights, Inc. v. Hampton* by a California United States District Court in which Judge Alfonso Zirpoli ruled in favor of the plaintiff (Hickerson), that the Commission cease ignoring the ruling in the *Norton* case, and that the Commission immediately:

> cease excluding or discharging from government service any homosexual person
> whom the Commission would deem unfit for government employment solely
> because the employment of such a person in the government service might bring
> that service into the type of public contempt which might reduce the government's

[86] *Bruce Scott v. John Macy et. al.,* June 16, 1965. https://casetext.com/case/scott-v-macy
[87] Ibid.
[88] Johnson, *Lavender Scare,* 202.

ability to perform the public business with the essential respect and confidence of the citizens which it serves.[89]

On December 21, 1973, in response to *Society for Individual Rights, Inc. v. Hampton,* the Civil Service Commission issues a bulletin to all federal agencies announcing they could no longer "find a person unvisitable for Federal employment merely because that person is a homosexual," but could only terminate or refuse to hire a person whose "homosexual conduct affects job fitness—excluding from such considerations, however, unsubstantiated conclusions concerning possible embarrassment to the Federal service."[90] Taken together, the rulings in *Scott v. Macy (1965), Norton v. Macy (1969),* and *Society for Individual Rights, Inc. v. Hampton (1973)* paved the way for a formal change in policy at the Civil Service Commission in 1975. The formal change in policy came on July 3, 1975 when the Civil Service Commission issued a press release announcing a "significant change from past policy—resulting from court decisions and injunction [sic]—provides applying the same standard in evaluating sexual conduct, whether heterosexual or homosexual."[91] This formal change in Civil Service Commission would not be the end of incidents of discrimination against homosexuals in the federal government but it did signify a major shift in federal policy.

Closing

The cruel injustices experienced by members of the LGBTQ+ community during the Lavender Scare are part of a painful chapter in our national history. Every effort was made during this historical investigation to be as thorough in research and objective in analysis as possible. The

---

[89] *Society for Individual Rights, Inc. v. Robert Hampton,* 63 F.R.D. 399 (N.D. Cal. 1973), October 31, 1973. https://casetext.com/case/society-for-individual-rights-inc-v-hampton-2
[90] Quoted in Lewis, "Lifting the Ban on Gays in the Civil Service: Federal Policy Toward Gay and Lesbian Employees Since the Cold War," 392.
[91] Ibid., 393.

analysis provided as part of this investigation is intended to provide as full a contextualization of the available evidence, as it concerns the relevance of James Webb both at the State Department and NASA. Again, I wish to acknowledge the historians, archivists, and librarians who have assisted with locating pertinent documentation in archives across the country and provided valuable insight into this history.

III. Research Methodology

- Attempted to locate and examine primary sources related to James Webb's time as Under Secretary of State (1949-1953) and the firing of homosexuals as well as his time as NASA Administrator (1961-1968) linking him directly to Lavender Scare or firing of Clifford Norton.

- Examined related secondary literature to establish historical context of James Webb's career, the Lavender Scare, and *Norton v. Macy (1969)*.

- This phase also included conversations with historians and archivists familiar with the context of both the Lavender Scare and James Webb's overall career.

- Using established context, attempted to locate further sources linking James Webb to the firing homosexuals in the federal work force during his time at the Department of State and NASA.

- Noteworthy evidence could take the form of memorandum to or from James Webb which directly linked him to actions taken during his time of service at either location. This evidence could include such as memoranda, reports, correspondence with key participants, notes, meeting minutes, or other documentation which established Webb's direct action.

- Hired contract historian to explore archival collections at the Records of the Department of State – National Archives, Archives II, College Park, Maryland.

- Presented findings of the research to the Office of the NASA Administrator.

Early in the process, limited access to important archival collections imposed by the COVID 19 pandemic presented a limitation to the depth of historical research in this investigation. Those collections (Archives II and Truman Presidential Library) remain closed until November 2021 (Archives II) and late spring 2022 (Truman Presidential Library). Until those collections were reopened, the preliminary investigation relied on up several secondary works of credible historians who have gone through those collections with similar research questions. The primary works consulted are listed in the bibliography of this report. While other works were consulted along the way, these works represent the primary relevant historiography.

With many important archival collections closed in early period of the investigation, we reached out to the archivists at National Archives II in College Park, Maryland who provided the following suggested "roadmap" for researching the James Webb Personal Papers at that facility.

The following is a research plan developed in consultation with NARA archivists for examining the records of the Department of State. Particular attention is given to the arrangement of those records and potential locations within those collections of relevant documentation.

The primary source for documentation on the Department of State, U.S. foreign policy, and events in various countries is the Department of State central files, part of RG 59: General Records of the Department of State.

From 1910 to 1963, the Department's central file is arranged according to a pre-determined decimal subject classification scheme known as the Central Decimal File. The file is broken into the following segments: 1910-29, 1930-39, 1940-44, 1945-49, 1950-54, 1955-59, 1960-63.

The central file for the period 1910 through 1949 is arranged subjectively in nine subject classes. Within these classes, the files are further broken down by subject:

♦Class 0: General. Miscellaneous

♦Class 1: Administration

♦Class 2: Extradition

♦Class 3: Protection of Interests

♦Class 4: Claims

♦Class 5: International Congresses and Conferences

♦Class 6: Commerce

♦Class 7: Political Relations of State

♦Class 8: Internal Affairs of States (This class is further divided into file categories on political affairs; public order, safety, health, and works; military affairs; naval affairs; social matters; economic matters; industrial matters; communication and transportation; navigation; and other internal affairs.)

Documentation created/reviewed by Webb is scattered throughout the files based on its subject. There is a small administrative file one the Under Secretary in Class 1 under file "111.16 WE".

The central file for the period from 1950 to January 1963, is arranged subjectively in ten subject classes. Within these classes, the files are further broken down by subject:

♦Class 0: Miscellaneous

♦Class 1: Administration

♦Class 2: Protection of Interests

♦Class 3: International Conferences, Congresses, Meetings and Organizations

♦Class 4: International Trade and Commerce

♦Class 5: International Informational and Educational Relations

♦Class 6: International Political Relations

♦Class 7: Internal Political and National Defense Affairs

♦Class 8: Internal Economic, Industrial, and Social Affairs

♦Class 9: Communications, Transportation, Science

The class number becomes the first digit in the file number.

Documentation created/reviewed by Webb is scattered throughout the files based on its subject. There is a small administrative file one the Under Secretary in Class 1 under file "110.12 WE".

Also in RG 59 are decentralized records of various high level, geographic, and functional offices of the Department. Those files can be a valuable supplement to the documentation found in the central files. For important information about the decentralized files see: https://www.archives.gov/research/foreign-policy/state-dept/rg-59-decentralized-files

The following files from the Executive Secretariat are likely to be of interest to this research. Finding aids are available in the Archives II research room and in the on-line Catalog:

RG 59 Entry A1-393. SUMMARIES OF THE SECRETARY'S DAILY MEETINGS. 1949 52. 10 in. Arranged chronologically. Summary memoranda of the proceedings of the Secretary's daily meetings. Each summary includes a list of the State Department officers meeting with the Secretary; the topics discussed; a brief summary of the discussions; and, for the year 1949, the names of individuals assigned to take action on subjects discussed. The summaries deal with both routine and administrative matters and with major crises of the period, such as the Korean conflict, the German problem, the Communists in China, the situation in Iran, the formation of NATO, the development of atomic energy, and Senator Joseph McCarthy's charges against the Department.

RG 59 Entry A1-394B. MEMORANDUMS OF THE SECRETARY AND UNDER SECRETARY. 1951 52. 4 in. Arranged chronologically. Chiefly copies of memorandums by and for the Secretary of State and Undersecretaries James E. Webb and David K. E. Bruce on a wide range of foreign policy, domestic political, and administrative matters. Most of the memorandums are signed by Special Assistant to the Secretary Lucius D. Battle and Jeffrey C. Kitchen of the Executive Secretariat's Policy Reports Staff. These documents consist of memorandums of telephone conversations, summaries of the Secretary's conversations with the President, reports of meetings, and notes regarding appointments and speaking engagements.

RG 59 Entry A1-395. AGENDA FOR THE UNDER SECRETARY'S MEETINGS. 1949 52. 5 in. Arranged chronologically in numerical sequence, UM A1 UM A448. Brief agendas that provide the date and time of each meeting and the topics scheduled for discussion. The meetings dealt with a wide range of subjects, such as military aid, interdepartmental cooperation, congressional hearings, psychological warfare, and the Department's position on legislative programs and on internal administrative matters. The meetings were usually attended by the division heads.

RG 59 Entry A1-396. INDEX TO RECORDS OF THE UNDER SECRETARY'S MEETINGS. 1949 52. 1/4 in. Arranged alphabetically by subject. A list, by subject, of the documents, action summaries, and minutes of the Under Secretary's meetings.

RG 59 Entry A1-396A. INDEX TO PROBLEMS CONSIDERED AT THE UNDER SECRETARY'S MEETINGS. Feb. 1949 Apr. 1949. 1/4 in. Arranged chronologically. A list of problems discussed, the action summaries and documents involved, decisions reached, and the names of persons given assignments.

RG 59 Entry A1-396B. POSITION PAPERS AND REPORTS OF THE UNDER SECRETARY'S MEETINGS. 1949 1952. 15 in. Arranged chronologically in numerical sequence, UM D1 UM D152. Documents introduced at the Under Secretary's meetings, including position papers, reports, and memorandums. Among the major topics covered are U.S. policy toward Asia, military aid to Latin America, and the situation in Guatemala.

RG 59 Entry A1-396C. MINUTES OF THE UNDER SECRETARY'S MEETINGS. Feb. 3, 1949 Jan. 25, 1952. 10 in. Arranged chronologically in numerical sequence (1 447). Summary memorandums of the discussions and actions taken at the Under Secretary's meetings. They are not verbatim accounts of the proceedings. Also included are lists of persons who attended each meeting.

RG 59 Entry A1-396D. ACTION SUMMARIES OF THE UNDER SECRETARY'S MEETINGS. Feb. 1949 Mar. 1951. 3 in. Arranged chronologically in numerical sequence, UM S1 UM S315. The action summaries provide the date and time of each meeting, the topics discussed, a brief summary of the actions taken, and a list of the documents presented.

RG 59 Entry A1-396E. NOTES ON THE UNDER SECRETARY'S MEETINGS. March 1951 Jan. 1952. 2 in. Arranged chronologically in numerical sequence, UM N321 UM N447. Similar to the action summaries, these notes on the Under Secretary's meetings include the date, time, actions taken, and a brief statement concerning the proceedings of each meeting. There are no notes for some meetings.

This roadmap of sources was critical to establishing where any potential evidence might be located that could connect James Webb's time at the Department of State with the Lavender Scare. When Archives II reopened for researcher appointments in November 2021, the historian contracted by the NASA History Office was able to examine the records of the United States Department of State; specifically the record groups listed below:

RG 59, 1945-49 Central Decimal File, File "111.16 We," (Box 450)

RG 59, 1950-54 Central Decimal File, File "110.12 We," (Box 430)

RG 59 Entry A1-393 Summaries of The Secretary's Daily Meetings. 1949 52, (Boxes 1-2)

RG 59 Entry A1-394b. Memorandums of The Secretary and Under Secretary. 1951 52, (Box 1)

RG 59 Entry A1-395. Agenda For the Under Secretary's Meetings. 1949 52, (Box 1)

RG 59 Entry A1-396a. Index to Problems Considered at The Under Secretary's Meetings. Feb. 1949 Apr. 1949, (Box 1)

RG 59 Entry A1-396b. Position Papers and Reports of The Under Secretary's Meetings. 1949 1952, (Boxes 1-3)

RG 59 Entry A1-396c. Minutes of The Under Secretary's Meetings. Feb. 3, 1949 Jan. 25, 1952, (Boxes 1-2)

RG 59 Entry A1-396d. Action Summaries of The Under Secretary's Meetings. Feb. 1949 Mar. 1951, (Boxes 1-2)

RG 59 Entry A1-396e. Notes on The Under Secretary's Meetings. March 1951 Jan. 1952, (Box 1)

RG 59 Entry A1-1194: Executive Secretariat/Correspondence Files with State Department Personnel, 1947-1953, (Boxes 24-26) 2

RG 59 Entry P-528: Office of The Legal Adviser/Records Relating to Loyalty and Security Issues, 1944-1954, (Boxes 11 -12)

RG 59 Entry A1-1187: Memoranda for The President, 1944-1951 (Boxes 1-7)

RG 59 Entry A1-1188: Secretary's Memoranda, 1949-1951 (Boxes 8-10)

RG 59 Entry A1-1189: Memoranda of Conversation, 1947-1952 (Boxes 11-14)

RG 59 Entry A1-1192: Records Pertaining to Appointments and Staff Meetings, 1947-1952 (Box 22).

The contract historian made five research trips into the National Archives at College Park, Maryland collections examining over 50,000 pages of documents covering the period from 1949-1953.[92]

Once the Truman Presidential Library reopened to researchers in the spring of 2022, an appointment was made by the acting NASA Chief Historian. From April 11-13, 2022, I [the acting NASA Chief Historian] conducted research on site at the Truman Presidential Library in Independence, Missouri[93]. At the Truman Presidential Library, I closely examined thousands of documents from the James E. Webb Papers[94], Dean G. Acheson Papers[95], Harry S. Truman Papers[96], Stephen J. Spingarn Papers[97], Charles S. Murphy Papers[98], and Donald S. Dawson Papers[99]. Again, my primary objective was to locate any evidence that might shed light on James Webb's relationship to the Lavender Scare. Notable examples would be any documentation/correspondence

---

[92] The National Archives at College Park, Maryland. https://www.archives.gov/college-park
[93] Harry S. Truman Presidential Library and Museum, Independence, Missouri. https://www.trumanlibrary.gov/
[94] James E. Webb Papers, Truman Presidential Library. https://www.trumanlibrary.gov/library/personal-papers/james-e-webb-papers
[95] Dean G. Acheson Papers, Truman Presidential Library. https://www.trumanlibrary.gov/library/personal-papers/dean-g-acheson-papers
[96] Harry S. Truman Papers, Truman Presidential Library. https://www.trumanlibrary.gov/library/truman-papers
[97] Stephen J. Spingarn Papers, Truman Presidential Library. https://www.trumanlibrary.gov/library/personal-papers/stephen-j-spingarn-papers
[98] Charles S. Murphy Papers, Truman Presidential Library. https://www.trumanlibrary.gov/library/personal-papers/charles-s-murphy-papers
[99] Donald S. Dawson Papers, Truman Presidential Library. https://www.trumanlibrary.gov/library/personal-papers/donald-s-dawson-papers

in which James Webb was either presented with actions on the firing of homosexual employees, any policy documents requesting his approval/review, or any reports in which additional information concerning the firing of homosexual employees was presented to James Webb. It is important to note that the later phase of the investigation (once archival collections reopened for research) revealed no new significant evidence related to either Webb's time at the Department of State or at NASA. Documentation located during that later phase did allow for great contextualization of previously available evidence including Carlisle Humelsine's packet of memoranda provided to Webb on June 24, 1950.

IV. Key Bibliography and Primary Sources

Adkins, Judith. "These People are Frightened to Death: Congressional Investigations and the Lavender Scare." Prologue Magazine, Vol. 48, No. 2 (Summer 2016).

Beemyn, Genny. A Queer Capital: A History of Gay Life in Washington D.C., New York and London: Routledge, 2015.

Edsall, Nicholas C. *Toward Stonewall: Homosexuality and Society in the Modern Western World.* Charlottesville: University of Virginia Press, 2003.

Fried, Richard M. *Nightmare in Red: The McCarthy Era in Perspective.* Oxford: Oxford University Press, 1990.

Johnson, David K. *The Lavender Scare: The Cold War Persecution of Gays and Lesbians in the Federal Government.* Chicago: University of Chicago Press, 2004.

Kirchick, James. *Secret City: The Hidden History of Gay Washington.* New York: Henry Holt, 2022.

Lambright, Henry. *Powering Apollo: James E. Webb of NASA (New Series in NASA History).* Baltimore: Johns Hopkins University Press, 2000.

Lewis, Gregory B. "Lifting the Ban on Gays in the Civil Service: Federal Policy Toward Gay and Lesbian Employees Since the Cold War." *Public administration review* 57, no. 5 (1997): 387–395.

Shibusawa, Naoko. "The Lavender Scare and Empire: Rethinking Cold War Antigay Politics,"

Diplomatic History, Vol. 36, No. 4 (September 2012).

Key Primary Sources:

In addition to archival collections at NASA Headquarters and NASA's Marshall Space Flight Center, documentation was also located via online collections at the National Archives and Records Administration (NARA).

The following documents represent the key available evidence utilized in this research. Individual documents are prefaced with a description which includes both title and location (file, series, and collection) of original documents held at by the National Archives.

The current analysis considered the work of other prominent historians who did have access to those collections prior to COVID-19. Prominently, the work of David Johnson closely examines the records held at the National Archives including:

- James E. Webb, "Meeting with the President, Thursday, June 22, 1950," Box 9, Entry 53D444, Secretary's Memoranda, 1949–1951, Records of the Executive Secretariat, RG 59, NARA.

- David D. Lloyd to Mr. Spingarn, July 3, 1950, in "Sex Perversion" folder, Box 32, WHCF (White House Central Files), HST (Harry S Truman) Library

Specific archival collections of interest in this investigation include:

- James E. Webb Papers, 1928-1980, Harry S. Truman Presidential Library (National Archives), https://www.trumanlibrary.gov/library/personal-papers/james-e-webb-papers

- Records of the Department of State – National Archives, Archives II, College Park, Maryland https://www.archives.gov/research/foreign-policy/state-dept/rg-59-central-files

Appendix I. Key Evidence

DEPARTMENT OF STATE
———
THE UNDER SECRETARY

June 28, 1950

S - The Secretary *I approve. H.A.*

    With the return to this country of Mr. Harriman, Mr. Dulles
and Mr. Bohlen, it seems to me that we should make plans for
the proper organization of our resources and thereby prevent
any confusion or hurt feelings.

    With respect to Mr. Dulles, it seems to me that he should
retain the same position in which we placed him when he came
in the Department; that is, shoulder to shoulder with Dean Rusk
as Rusk's high level adviser in matters affecting the Far East
and giving all possible assistance to Rusk. In this way Rusk
could bring him in on those high level conferences where he is
needed. In addition, I think Mr. Dulles should be available to
pass on to Mr. Hickerson any ideas he has about UN matters or
for such advice as Hickerson desires from him.

    With respect to Mr. Bohlen, I suggest that he be attached
to the Policy Planning Staff and take his leadership and guidance
from Mr. Kennan. In that way Mr. Kennan can direct his activities
and bring him into such meetings as he feels are appropriate.

    With respect to Mr. Harriman, I suppose the President will
wish to install him in his new position at the White House and
may wish to designate him as our point of contact at the
White House in handling the fast moving operations of the immediate future. If the President desires this it will be necessary
to have an understanding as to how we handle our liaison with
Lay, Elsey, Murphy and others who have been involved in various
phases of our White House clearances. I suggest this be left to
Harriman to work out with his colleagues at the White House.

    My own plans are subject to change at any time, but I thought
I would take my family on to North Carolina tomorrow morning,
which means I would arrive at Kitty Hawk late in the afternoon.
I will then be available to return here either that evening or at
any time you wish me back. I have arranged with McWilliams to
have me flown back and it would be possible for me to get here in

James E. Webb to Secretary Dean Acheson, June 28, 1950, James E. Webb Papers Box 24, Notes on
Conversation with Secretary of State Organization and Administration of the Department of State
Folder, Truman Presidential Library.

a few hours, if need be. Arrangements have been made for the
Secretariat to channel to Matthews all matters which I would
normally handle, except those which must come to you. Matthews
will be the action officer on these and will be authorized to
affix any action signature which may be required.

James E. Webb

U:JW:HED.

July 5, 1950

Memorandum for the Hoey Subcommittee Sex
Pervert Investigation File

I went to the Capitol today to confer with Senator
Hoey about this investigation.

I told him that, as he had requested at our previous
meeting of June 28 (at which Charlie Murphy and Jim Webb were
present), we had considered the question of whether the entire
Subcommittee hearings should be held in executive session,
including the medical testimony, or whether the medical testimony might be given at public hearings with the rest in executive
session.

I told him that it was the unanimous opinion of the
White House staff, including Matt Connelly, Charlie Murphy,
Don Dawson, and others, as well as that of Surgeon General
Scheele, that all of the hearings, including the medical
testimony, should be in executive session. I went over the
reasons for this briefly. The Senator said he was glad to
get our views and indicated he would try to work it out that
way, although there might be some opposition from the Republicans, particularly Senator Mundt.

The Senator also told me that he proposed to get
the records of the Washington Police on arrests and convictions
in this field and then query the various agencies as to what,
if anything, they had done about individual employees involved.
This may present some problems on the disclosure of information, but
I do not see how it can be avoided since these police records are
public records and are not within the President's directives
about non-disclosure of personnel files and information.

The Senator said that Superintendent Barrett of the
Metropolitan Police had called him this morning at Matt Connelly's
suggestion and had offered to cooperate fully in connection with
the investigation.

I also made two further suggestions to Senator Hoey,
both of which were made to me by Dr. Scheele. The first was
that a Medical Staff Adviser be appointed by the Subcommittee
for the purpose of its hearings. Scheele recommended for this
purpose Dr. Tex Buxton, President of the Washington, D.C.,
Psychiatric Association. I know Dr. Buxton and think he would
be a good man for this purpose and therefore gave his name to
the Senator.

- 2 -

The other suggestion was that Navy and Army security officers (presumably the ONI Director and the Army Director of Intelligence) be the first security officers called by the Subcommittee after the medical testimony is in. Dr. Scheele made this suggestion because the military security people have had particular experience with the homosexual problem on a pretty big scale. Senator Hoey seemed to react favorably to both of these suggestions.

He said that he expected to call his Subcommittee together within the next few days to map out their course of direction.

S.J.S.

RESTRICTED

July 10, 1950

Memorandum for the Hoey Subcommittee Sex
Pervert Investigation File

Mr. Francis Flanagan, Chief Counsel of the Hoey Subcommittee, came in to see me Saturday morning, July 8, and
spent about an hour and a half discussing the investigation
with me. He said that Senator Hoey had asked him to do so.

I went over the same ground with him that Mr. Webb,
Mr. Murphy and I had gone over with Mr. Hoey.

Mr. Flanagan apparently intends to start off with
the security officer testimony rather than the medical testimony.
He said he had talked to a lot of the doctors and that he did
not think they had a practical approach to the matter. He said
they talked in terms of a large percentage of the male population
having homosexual tendencies, whereas he was thinking only in
terms of overt acts. I tried to give him the medical picture
as I saw it, but I'm afraid I was not very successful.

He told me that from preliminary conversations with
the security people they were not able to produce much dope in
documented instances in which homosexualism had endangered
security, except that CIA had some World War I instances of this
although apparently in other countries. Despite the lack of
documentation he seemed convinced that homosexualism represents
a serious security threat and my suggestion that it should be
squared up against other types of security threats by individuals
resulting from normal sexual or non-sexual activity did not seem
to impress him much.

He raised the question of getting names and files from
the agencies. I told him what we had told Senator Hoey -- that
this would put the matter squarely in the President's lap and we
hoped this would not be necessary, that we would have no objection
to the agencies furnishing statistics on the matter, which he is
in fact already collecting. He said that Senator Mundt or other
minority members of the Committee might be insistent on this.
He himself did not see how the Subcommittee could determine
whether the existing method of handling this problem was ade-
quate unless they could look at some files on a sampling basis.

In any event, they are getting the police records of
the Washington Police on arrests and convictions in this field
and will make a check to determine how many of the people in
this group are Government employees or have been. If any of
these are employed by the Government, this would seem to require that the employing agency report to the Subcommittee how
it has handled the particular case.

- 2 -

As far as the sampling look at the files is concerned, I suggested only as my personal opinion the possibility that the agencies might furnish memoranda summarizing individual files without mentioning any names. This would give the Subcommittee the necessary information as to how the determination of homosexualism had been made and what had been done about the matter. Flanagan said that would probably satisfy him, but he seemed doubtful that it would satisfy Senator Mundt. He suggested that Mundt might say that the agencies had not given an accurate account of what was in the files (which is, of course, what Senator McCarthy said about the loyalty files).

Flanagan said he would keep in touch with me in connection with the investigation. I have the impression that we may have some difficulties from his direction because he seems to have pre-judged one of the central issues that the Subcommittee has to decide, namely, how serious the threat is of the homosexual employees, particularly in relationship to other types of security threats. Flanagan seems strongly committed to the position that the homosexual is the most serious security threat of all, and he seems to regard the fact that there is scant documentation of this as an unfortunate accident. On the basis of my experience as a counter espionage and security officer during the war, I personally am of the opinion (I believe I can cite chapter and verse to support it) that other types of security threats are more dangerous than the homosexual, although no doubt he represents one type.

S.J.S.

cc: Mr. Murphy
Mr. Dawson
Mr. Maletz
Mr. Elsey - Mr. Hechler - Mr. Lloyd -
   Mr. Bell - Mr. Nash - Mr. Neustadt

DEPARTMENT OF STATE

DEPUTY UNDER SECRETARY

June 24, 1950

MEMORANDUM FOR MR. WEBB

Subject: Department's participation in the homosexual
          inquiry by the Hoey Committee

I am attaching hereto several memoranda.

These are:

    (1) A background paper on the problem of homosexuals
        and sex perverts in the Department of State.

    (2) A report of a meeting between Mr. Flannagan,
        Senate Investigations Staff, and Mr. Arch Jean,
        Chief of Departmental Personnel.

    (3) A memorandum suggesting a basis for discussion
        and briefing for your meeting with Senator Hoey
        on the objectives and methods of operation of the
        Senate Committee established to look into the
        problem of homosexuals and moral perverts in the
        Federal Government.

    (4) A memorandum suggesting the organization and
        principles to govern the Department's participa-
        tion in the Senate inquiry.

    (5) A list of the Senate Committee.

Carlisle H. Humelsine

A CHH:jgc

DEPARTMENT OF STATE

DEPUTY UNDER SECRETARY

TO: U - Mr. Webb

FROM: A - Carlisle H. Humelsine

SUBJECT: Problem of Homosexuals and Sex Perverts in the
Department of State.

Homosexuality, which is sexual attraction to a person of the
same sex, is as old as the history of mankind. From time immemorial
all races of man have had to deal with the subject. Some have condoned
it and some have condemned it. Studies have been made which purport
to relate the strong rise of homosexuality with the accompanying
decline of the Egyptian, Greek and Roman Empires. Some experts hold
that where the mores of a people have condoned homosexuality through
apathy, the vigor and virility of that people have been emasculated,
and that where the homosexuality of an individual has been established
in a society where modesty demands concealment, the position of that
individual has been weakened psychologically and sociologically.

Many of the men who have studied homosexuality tell us that
homosexuals are neurotic, characterized by emotional instability,
that they represent a type of regression to man's primitive instincts
and that they live a life of flight from their inversion and of fear of
detection. They are content and at ease only when surrounded with other
homosexuals. They meet at known homosexual gathering places, seek each
other in cocktail lounges and public parks, and rarely live with anyone
other than another homosexual. They come from all walks of life and all
strata of society. They often disassociate themselves with their early
childhood and family connections and endeavor to build a pseudo-cultural
background around them. Many of them develop strong hate fixations
which often colors and affects their thinking and behavior. These
fixations may be on the mother, father, a brother or sister, or on all
members of the opposite sex.

Until very recent years the Department of State, as well as the
several agencies of the Federal Government, tolerated homosexuals in
its employment solely because not much was known about them or who
they were. Occasionally when one was found he was dismissed or
reassigned, depending upon the circumstances surrounding the individual
case. It was the type of problem that most officers of the Federal
Government, not conversant in the subject, would rather not consider.
It therefore was allowed to exist and to grow.

It was

CONFIDENTIAL

It was not until January 1947 when Mr. Peurifoy became Assistant Secretary for Administration that the problem of homosexuality in the Department of State was dealt with in a direct and forthright manner. It came about through the investigation of a homosexual which lead our investigators to other homosexuals in the Department, which in turn enabled investigators to discover still others on the Department rolls. With this knowledge it was determined that there probably were a number of such people on the rolls. Since Civil Service rules preclude the appointment of anyone who is guilty of "criminal, infamous, dishonest, immoral or notoriously disgraceful conduct", the Department concluded that it was within its power to separate an individual who was found through investigation to be homosexual. The same reasoning was adopted with respect to the Foreign Service.

Our investigations and studies of the subject revealed that homosexuals are, generally speaking, undesirable as employees for a number of reasons: (1) They create a morale problem, i.e., most men who are considered by the majority of us to be normal desire not to work or associate with homosexuals; (2) They are emotionally unstable, i.e., many of them have told our investigators of the inexorable pain and humiliation they would suffer if exposed to family and friends, and some have even threatened suicide; (3) Usually they live in a world all to themselves associating and consorting with other homosexuals; (4) They indulge in acts of perversion which are legion and which are abhorent and repugnant to the folkways and mores of our American society; (5) They are immoral in their sexual behavior seeking sexual gratification from one person one night and from another person the next in a paltry and endless gesture at a happiness they never realize.

Why homosexuals have been employed in the Department of State is a question in which we have been profoundly interested. It has been found that many of them leave their family and childhood surroundings in an attempt to create a pseudo-cultural background around them. Many of them are therefore attracted to the Department of State because of its cultural atmosphere and attainments, both in the Department and the Foreign Service. We have found that most of those discovered in the Department hope for a career in the Foreign Service. Many of them have told our investigators that they believe the chances of detection in a foreign country are far less than in this country. It is known that some of them attract other homosexual friends into the service. We are aware of this possibility and do our best to prevent it.

The Department determines whether a person is a homosexual or sex pervert through the media of investigation. There are two investigators on the staff of the Security Division who devote full

time to

time to the detection of such individuals and the study of the problem.
There are several cases under consideration at the moment. When
information or evidence is received that an employee is suspected of
being a homosexual, an investigation is assigned to one of these two
investigators. A thorough and comprehensive inquiry into the matter
is made to ascertain all the facts in the case, bearing in mind the
peculiar susceptibility of such cases to possible malicious charges.
The investigation entails inquiries at all places of employment, all
residences and habitats. The investigation also attempts to determine
with whom the person associates and whether any of his friends or
associates is homosexual. All available records, including school,
credit, police and other investigative agency records are checked.
All character references and other people who may know the subject
of the investigation are interviewed personally. If the circumstances
warrant it, he may be placed under surveillance to determine whether
he frequents known homosexual places or associates with other known
homosexuals. In all cases the person under investigation is accorded
a personal interview not only by the investigator but often by the
Chief of either the Division of Departmental Personnel or Foreign
Service Personnel, depending upon the service in which he is employed.
If the person is determined to be a homosexual through investigation
or admission, he is promptly separated from the Department.

The human element of the problem has always caused us considerable
concern and has been made more difficult of resolution because the
medical profession itself is at such sharp variance as to the cause
and the possibility of cure of homosexuality. One school of thought
holds to the theory that homosexuality is congenital. Other schools
hold that it is acquired, while a great number admit that evidence
is lacking that it is either. Some, especially in the psychiatric
field, contend that homosexuals can be cured while others who have
studied the problem maintain that there is no cure.

We believe that most homosexuals are weak, unstable and fickle
people who fear detection and who are therefore susceptible to the
wanton designs of others.

We have no evidence, however, that these designs of others have
caused a breach of the security of the Department. Yet the tendency
toward character weaknesses has led us to the conclusion that the known
homosexual is unsuited for employment in the Department.

Memo Packet Page 5
Memo from Arch Jean to John Peurifoy
Confidential Files (Truman Administration), 1938-1953)
Sex Perversion [Investigation of Federal employees]

CONFIDENTIAL

(2)

STANDARD FORM NO. 64

Office Memorandum · UNITED STATES GOVERNMENT

TO   :   A - Mr. Peurifoy   DATE: June 20, 1950

FROM   :   DP - Arch K. Jean

SUBJECT: Report of meeting with Mr. Flannagan, Senate Investigations Staff

As you requested, I met with Mr. Flannagan of the Senate
Investigations Staff this morning and found him to be a personable
individual who knows that he has a disagreeable job to perform and
is searching for ways and means to accomplish his task without fanfare and without embarrassment to the agencies or the people involved.
We talked for approximately an hour and a half, more or less at random,
so it is difficult to relate accurately the conversation. Nonetheless,
here are the important matters that were discussed. They are not given
in their order of importance necessarily.

1. He attempted to rationalize his position with respect to
    release of agency files to the Committee. He stated
    definitely that we would be formally requested to give our
    files to the Committee for such use as they may deem proper
    and necessary. I told him that I believe the President's
    order on release of confidential personnel information would
    preclude our complying with such a request, but that in the
    final analysis only the White House could make that determination.
    It was his view that unless the files were released to the
    Committee the investigation would reduce itself to a fiasco,
    and in such event, the Department of State specifically would
    suffer in the eyes of the public. I expressed no opinion of
    my own on this point, other than to say that I could foresee
    the possibility of their conducting a meaningful investigation
    without the use of the investigation files. It seems to me
    that names of individuals and circumstances surrounding their
    cases would not necessarily help them in determining a procedure
    to be followed by all agencies in the handling of the problem.

2. Mr. Flannagan stated that we should be prepared to state our
    views with respect to the security risk involved in the employ-
    ment of a homosexual. Likewise we should be prepared to tell
    the Committee how we view homosexuals from the sociological
    standpoint.

3. Flannagan asked me who in the Department I would consider to
    be well informed on the subject and therefore who the Committee
    might call to testify. In this connection he stated that he
    was aware of the part Finlator*has played. In answer to this
    question I told him that in my personal opinion you, as well
    as Sam,** Pete, Don, Don Smith, and myself are all conversant
    and of like mind with regard to the subject.

* Finlator is an investigator who has been working on this problem   4. He asked

** Mr. Boykin - head of Office of Controls
    Mr. Martin .. .. Personnel
    Mr. Smith   .. .. Foreign Service Personnel
    Mr. Nicholson .. .. Security Division

CONFIDENTIAL

DECLASSIFIED.
E.O. 12065, Sec. 3-402
State Dept. Guidelines, June 12, 1979
By NLT [illegible] NARS, Date 5/10/87

Memo Packet Page 6
Memo from Arch Jean to John Peurifoy
Confidential Files (Truman Administration), 1938-1953)
Sex Perversion [Investigation of Federal employees]

CONFIDENTIAL

-2-

4. He asked whether or not the following statistical data
could be supplied him at an early date: Since July 1, 1945,
(if a later date would save the Department a great deal of
research and effort, he would agree to another point of
departure) how many employees were allowed to resign, how
many were dismissed, and how many suspects do we now have
under investigation. With respect to those who resigned
or were dismissed, how many were suspects and how many
were admitted homosexuals.

5. He asked about the procedure we followed from the point of
first knowledge or suspicion to the conclusion of the case.
I described the procedure stressing that the decision was
always an administrative decision though sometimes related
to security.

6. He also asked if we might be able to cite examples of homosexuals' tendency to locate employment with others of their
kind in the same agencies. I told him that we have first-
hand knowledge of the fact that such a tendency exists and,
as a matter of fact, it has lead us to cases which we were
not aware of.

Mr. Flannagan discussed at some length the desires of the Committee
and of the Staff to keep the investigation on a high plane and to conduct
most, if not all, of the hearings in executive session. He thought it
might be necessary to hold at least one session in public to satisfy
some of the more politically minded members of the Committee. He was
hoping, nevertheless, to convince the Committee that such would not be
desirable. He also described the lengths one of his staff members was
going to in developing the medical side of the problem.

All in all, I was convinced of his sincerity to conduct an intelligent,
non-political investigation and I told him that I was sure he could count
on the State Department's cooperation. I got the impression, however, that
Flannagan has already concluded that homosexuals should not be employed
in government under any circumstances and that doubt should always be
resolved in favor of the government.

It was my opinion that we should supply the statistical data requested.
I believe our only meaningful data, however, would date from January 1947.
I promised to let him know shortly whether or not we were going to comply
with his request.

cc: CON - Mr. Boykin
    FP - Mr. Smith
    PER - Mr. Martin
    SY - Mr. Nicholson

CONFIDENTIAL

~~CONFIDENTIAL~~

Suggestions as to the objectives of the Committee and Methods
of operation

I. Objectives of the Committee

1. The Committee should undertake to evaluate the problem in its
totality and should not undertake to determine the innocence
or guilt of individual employees.

2. The Committee should focus on the study of present conditions
as they pertain to policies and procedures in the several
Agencies and should steer clear of digging up individual cases
which have been handled in the past and which have no bearing on
present conditions or a proper course of future action.

3. The Committee should seek to answer the following specific
questions:

a. What specific administrative, security and other problems,
, present or potential, are posed by homosexuals in the
Federal Government?

1. Is a homosexual or a moral pervert a security risk?

2. If considered a security risk in a sensitive agency,
should they be employed in a non-sensitive agency?

3. If not considered a security risk, should homosexuals
or moral perverts be employed in the Federal Government
as a matter of policy?

4. What action is dictated by the best medical judgment?

II. Suggestions for Committee Procedure

1. The Committee should conduct its investigation on the highest
possible plane, free of partisan thinking and action.

2. All business of the Committee should be transacted in executive
session.

3. The Committee should obtain from a competent medical board all
pertinent medical data available on the question, together with
recommendations of this board for dealing with the problem.

UNCLASSIFIED

Signature   05-11-42   ~~CONFIDENTIAL~~
                  Date

4. The Committee should ask the head of each Department and Agency of the Government to designate a responsible officer of his Department or Agency who will be the official liaison for the Department or Agency and with whom the Committee will conduct its business. Of course, in exceptional circumstances the Committee should take unto itself direct liaison with individual employees of the several Departments and Agencies.

5. The very magnitude of the problem precludes the Committee from effectively reviewing individual cases or actions of the several Departments and Agencies with respect to these cases. It should, therefore, not call homosexuals or moral perverts before the Committee. Neither should it undertake to judge individual cases. Nor should the Committee request or expect to receive from the several Departments and Agencies names of known or suspected homosexuals or moral perverts presently or previously employed; nor should it request or expect to receive investigation reports and other confidential information concerning individuals presently or previously employed. It is believed that Departments and Agencies could not release such information without the approval of the White House.

6. When the Committee has completed its study of the problem and before making a report to the Congress and American people, it should request the Administration to recommend administrative procedures and machinery needed to carry out the basic changes called for by the Committee's findings.

Organization and principles to govern the Department's
participation in the Committee inquiry.

The following steps should be taken to enable the Department to
effectively handle its relations and activities with the Hoey Committee.

1. The Deputy Under Secretary for Administration or his Deputy
    should be named as the Department's spokesman for dealing with
    the Senate Committee, with individual members of Congress
    and with the press. All actions and pronouncements of the
    Department relating to this subject should be exercised by
    him or under his personal direction. The Secretary and Under
    Secretary should be kept informed of all significant
    developments and should be available for behind the scene
    activities, when necessary.

2. There should be an Ad Hoc Committee at the immediate disposal
    of the Deputy Under Secretary (or his Deputy) to serve as
    his sounding board and to advise him on courses of action
    under varying circumstances. This Committee should be composed
    of Mr. Fisher (L), Mr. Boykin (CON), Mr. Martin (PER),
    Mr. Player (P), and Mr. Horace Smith (R).

3. That the Department agree to provide the Committee with such
    statistics as will be useful to that Committee without
    jeopardizing the Department's personnel and security programs.

4. The Department will resist any attempt of the Committee to
    obtain names of individuals and files.

UNCLASSIFIED

NLT·HL   5·10·72
_____   _____
Signature   Date

Members of Sub-Committee on Homosexuals
and Moral Perverts

Clyde R. Hoey   North Carolina

Herbert R. O'Conor   Maryland

James O. Eastland   Mississippi

John L. McClellan   Arkansas

Karl E. Mundt   South Dakota

Margaret Chase Smith   Maine

Andrew F. Schoeppel   Kansas

6/22/50

*Sex Pervert
File
6/26/50*

QUALIFIED MEDICAL WITNESSES

Dr. Leonard A. Scheele, Surgeon General, U.S. Public Health Service

General William Menninger, Topeka, Kansas

Karl Menninger (brother of General)

Captain George Raines, USN, Chief of Psychiatry, Bethesda Naval
    Hospital

Colonel Inwood, USA, Head of Psychiatry, Walter Reed Hospital

Colonel John Caldwell, USA

Commander Thomas Harris, USN, Chief of Psychiatry, Navy Bureau
    of Medicine

Dr. Robert Felix, Head of Psychiatry, U.S. Public Health Service

Dr. Lawrence Kubie, N.Y.C.; distinguished writer on psychiatric matters

Dr. Robert Knight, Head of Riggs Foundation, Stockbridge, Mass.;
    President of American Psychoanalytic Association

Dr. Leo Bartemeier, Detroit; President, International Psychoanalytic
    Association

Dr. Rex Buxton, Washington, D.C.; President, Washington Psychiatric
    Association.

Revised

THE WHITE HOUSE
WASHINGTON

June 30, 1950

NOTE FOR MR. DAWSON

    I would very much appreciate
your reaction about the question of
public hearing vs. executive session
mentioned in the attached memorandum.

                  S. J. S.
                  S.J.S.

Note from Stephen J. Spingarn to Donald S. Dawson with Attached Memorandum
National Archives Identifier: 54538201
Creator: President (1945-1953 : Truman). Office of the President. 4/1945-1/20/1953
From: File Unit: Sex Perversion [investigations of Federal employees], 1945 - 1953
Series: Confidential Subject Files, 1945 - 1953
Collection: Confidential Files (Truman Administration), 1938 – 1953

June 29, 1950

MEMORANDUM FOR THE HOEY SUBCOMMITTEE SEX PERVERT INVESTIGATION FILE

Yesterday afternoon Jim Webb of State, Charlie Murphy and I went up to see Senator Hoey about this matter, at his request.

We spent over an hour discussing the whole situation and a most useful interchange of views took place. Mr. Webb gave the Senator some material on the subject which Humelsine of State had prepared. I also gave the Senator some background material on the subject and a list of qualified medical witnesses which I had prepared on the basis of advice from Surgeon General Scheele and others.

We suggested that the hearings begin with testimony by competent medical authorities on the nature and scope of the problem, this testimony being designed to put the problem in proper perspective. After that could come testimony from senior Government security officers about the security problems involved. We noted in this connection that homosexuals were one category of security risks, and we discussed some of the other categories.

The Senator seemed to be very receptive to the ideas advanced about the hearings. The question came up about the Subcommittee requesting Government agencies for names and files of suspected or actual homosexual employees. Senator Hoey said that he had talked to Peyton Ford about the matter in terms of getting the statistics on the situation rather than names and files. Peyton Ford had said that Justice would collect this information for the Subcommittee. Senator Hoey thought it would be best if the Subcommittee collected it directly or possibly through the Civil Service Commission. Mr. Murphy agreed with this viewpoint.

Mr. Murphy expressed the hope that the Subcommittee would not find it necessary to call on the agencies for names and files. He said that, on the basis of the 1948 Presidential directives, the agencies would have to decline and refer the matter to the White House which would put it right in the President's lap. Mr. Murphy hoped this could be avoided. The Senator indicated that he shared that hope although he could not, of course, be certain what his Subcommittee would do. He indicated that it was a dirty job which he had not wanted but that he was going to do his best to do it right, and in a quiet and unspectacular way. I was impressed by his straightforwardness and sincerity about the whole matter.

The Senator asked our opinion as to whether any part of the hearings should be public. He apparently wants to state in advance how the hearings will be conducted and not wobble back and forth between public hearings and executive sessions according to the pressure of the moment as the Tydings Subcommittee has done. He thought that the medical testimony at the beginning might be public and the rest in executive session. We were of two minds about it. Mr. Murphy's reaction was that it would be best to have the whole hearing in executive session. Jim Webb was not certain and I was inclined to believe that the medical testimony should be public and the rest in executive session. The Senator asked us to think about it some more and get in touch with him. It was agreed that I would act as liaison man with him.

I talked to Peyton Ford today and told him of our visit with Senator Hoey and also asked his views about the public hearing question. Peyton was rather strongly inclined to the view that the medical testimony should be public.

S.J.S.

7/7/50

Mr. D:

What was it that you were going

to send Henry Hubbard about homos???

js

our policy
letter of aug.
must you see
Personnel —

Run off a 100 and
give to HH

Mr. D talked to Henry Hubbard +
it was decided Hubbard should
contact agencies by phone —
100 copies of 8/5/48 letter
Sent to Hubbard to be
distributed on
request.

Memorandum from Stephen J. Spingarn to Donald S. Dawson with Note
National Archives Identifier: 54538200
Creator: President (1945-1953 : Truman). Office of the President. 4/1945-1/20/1953
From: File Unit: Sex Perversion [investigations of Federal employees], 1945 - 1953
Series: Confidential Subject Files, 1945 - 1953
Collection: Confidential Files (Truman Administration), 1938 – 1953

THE WHITE HOUSE
WASHINGTON

June 29, 1950

MEMORANDUM FOR MR. DAWSON

Subject: Conversation with Senator Hoey about his Committee
           Counsel's request to several agencies for names and
           file information about known or suspected homosexuals.

        I talked to Senator Hoey about this matter this after-
noon. I told him about the letter which his Committee Counsel,
Mr. Flanagan, had sent to Secretary Sawyer on June 15 as well as
a similar request, oral or in writing, to other agencies.

        I told the Senator that in view of the conversation
which Jim Webb, Charlie Murphy and I had had with him yesterday
it was our understanding that we could tell the agencies to
disregard these requests from Mr. Flanagan for the time being
at least and until the Hoey Subcommittee had met and established
its procedures and policy.

        Senator Hoey said my understanding was entirely right
and that we could proceed on that basis.

        I return herewith the letter to you from Secretary Sawyer
and its attachments.

        Just to be on the safe side, it occurs to me that it
would be a good idea to make sure that all the principal agencies
at least understand this situation. It occurs to me that it is
possible that some of them may have received previous requests
from Mr. Flanagan which we have not heard about and may be taking
steps to honor these requests without consulting the White House,
perhaps unaware of the fact that the matter comes within the purview of the President's directives of March 13 and August 5, 1948
dealing with non-disclosure of file information to Congressional
committees and others.

                                        S.J.S.
                                        S.J.S.

Meeting Minutes-under secretary's meeting October 17, 1949
RG 59 Entry A1-396C Minutes of the Under Secretary's Meetings UM Minutes Feb.
3. 1949-Jan.25, 1952 Box 1

SECRET

DECLASSIFIED
Authority 957300

-6-

Mr. Acheson indicated his general satisfaction with the reports noted above. He indicated a desire to accept Mr. Webb's invitation to meet with the group once a month or as often as it seems desirable. At the opening of the meeting he had expressed his satisfaction with the progress that had been made with the new first team and expressed confidence in the top echelon. He also informed the group of his division of labor with Mr. Webb, in which the latter would take on organization and current business, leaving him free to work with the Policy Planning Staff and other policy groups.

CC - Mr. McWilliams, S/S
      Mr. Barnes, S/S-R
      Mr. Brown, S/S
      Mr. Smith, G
      Mr. Wilgus, U

~~RESTRICTED~~   February 7, 1950

To: S - The Secretary
Through: S/S

From: H - Mr. McFall

Subject: Testimony before Congressional Committees

At a recent meeting in your office it was suggested that policy matters should be discussed before Congressional Committees only by the senior officers of the Department. I was asked to prepare for your approval a proposed procedure on this subject.

Attached as Tab B you will find a procedure which would govern testimony by Departmental officers. This procedure would limit the authority to testify on policy matters to the Secretary, the Under Secretary, the Counselor, the Legal Adviser and the Assistant Secretaries. It would also authorize the Deputy to any of the aforementioned officers to so testify when he is "Acting." I think this is the logical place to draw the line. To limit it to the Secretary and Under Secretary would impose an almost intolerable burden, and to extend it below the Assistant Secretary level would so broaden it as to defeat the purpose of the limitation. Authorization for a Deputy to testify is necessitated by frequent absences from the Department by the Senior Officers.

Since this matter is one that will require the cooperation of the Congressional Committees, I believe that it would be advisable for Mr. Peurifoy and me to discuss it with the Foreign Relations and Foreign Affairs Committees and with the Appropriations Committees before it is put into effect. In the event that all the Committees agree to the procedure, it would then be published as a Departmental Regulation.

Recommendations: (a) That you approve the attached procedure.

                  (b) That Mr. Peurifoy and I be authorized to
                      discuss the procedure with the aforemen-
                      tioned Congressional Committees and seek
                      to obtain their agreement.

                  (c) That the principles in Tab B be incor-
                      porated in the present regulation (231.1)
                      governing appearance before Congressional
                      Committees.

~~RESTRICTED~~

Memorandum—under secretary's meeting Feb. 7, 1950
RG 59 Entry A1-396B Position Papers and Reports of the Under Secretary's Meetings
1949-1952, Box 1

Testimony of Departmental Officers Before Congressional Committees

The following principles will govern testimony by Departmental personnel before Congressional Committees:

1. Any matter involving the determination of the policy of the President and the Department will only be discussed before Congressional Committees by officers of the Department who are appointed to their positions in the Department by the President, i.e., the Secretary, the Under Secretary, the Counselor, the Legal Adviser and the Assistant Secretaries. The only exception which will be made will be in the case where a deputy or another officer is "acting" during the absence of one of the aforementioned officers.

2. Other officers of the Department testifying before Congressional Committees, will, in the event they are questioned on a matter involving the formulation of policy, inform the Committee that under Departmental Regulations such a question may only be answered by a senior officer of the Department. An offer to have a senior officer of the Department appear before the Committee, in the event the Committee so desires, should be made, unless a senior officer is present or scheduled to appear.

3. These regulations will be judiciously applied by Departmental officers, and invoked only with respect to matters clearly involving the formulation of policy. Questions involving the existence or application of established policy may be dealt with, with due regard to security considerations, in accordance with the present regulation.

Meeting Minutes-under secretary's meeting March 1, 1950
RG 59 Entry A1-396C General Records of the Under Secretary's Meetings UM Minutes Feb.3, 1949-
Jan. 25, 1952 Box 1

S E C R E T (21)

DECLASSIFIED
Authority NND 953300

MEMORANDUM FOR FILE

Under Secretary's Meeting March 1, 1950

1. Security Presentation

Mr. Humelsine will consider, in cooperation with the Assistant
Secretaries, the feasibility of having similar security presentations to the members of each of the Assistant Secretary's offices
in order that the personnel of the Department may not only have the
feeling that the Department is secure, but also that they may understand clearly the procedures which are designed to protect the
employees' rights and interests.

2. Questions Relating to the Briefing on Southeast Asia

A question was raised by Mr. Humelsine, who said that in his
judgment (a) seven people in Indochina were not enough, (b) he doubted
their competence, and (c) in summary, why, if we knew it was a hot spot
did we not do a better job of beefing up our manpower in that area.

Mr. Merchant explained that we had not because it was a guerrilla
warfare area, Secretary Johnson was over economical with respect to
providing military attaches, and the French were peevish about bringing in more of our people.

Mr. Humelsine offered complete support from A for beefing up
this area.

Mr. Rusk made the point that we must do a better job of identifying points of infection and then putting resources at these points.

Mr. Webb took the occasion to point out the responsibility of
the geographic and functional areas. He said on the one hand FE, for
example, has the responsibility for estimating its needs and to go to
the administrative area to get those fulfilled. On the other hand,
he believes the administrative area has the responsibility to take
the initiative with FE, to know about our global position, and to
assist the Secretary in our deployment of our resources.

It was in this discussion that Mr. O'Gara expressed his concern
about reorganization developments; in particular he referred to the
compartmentalization of the geographic bureaus as the result of the
executive office presidential theory. Mr. Webb acknowledged that he

was

S E C R E T

CONFIDENTIAL

UM S-164

UNDER SECRETARY'S MEETING

Action Summary

10:00 A.M., Friday, March 10, 1950
Room 5104

Security Question

1. Mr. Humelsine announced that officers receiving inquiries respecting persons listed in the McCarthy accusations should reply with confidence that there is positive evidence that these security charges are unfounded.

2. Mr. Peurifoy's office will consider the presentation of our security system and procedures to a public information conference which the P area will hold in the near future.

3. Mr. Peurifoy's office will also look into the possibility of some similar presentation through magazines at the appropriate time.

4. Mr. Humelsine stated that security presentations to the personnel in the Department would be scheduled after the presentation on the Hill and a background presentation to the press.

DECLASSIFIED

By _____ NARS, Date 4-3-79

CONFIDENTIAL

Meeting Minutes under secretary's meeting April 19, 1950
RG 59 Entry A1-396C General Records of the Under Secretary's Meetings, UM
Minutes Feb. 3, 1949-Jan. 25, 1952 Box 1

S E C R E T

- 2 -

5. Bipartisan Foreign Policy

It was made clear that while Mr. McFall will be the pivotal point in the Department for the coordination of bipartisan foreign policy efforts, Mr. Webb will continue to follow these activities very closely. Mr. Webb pointed out that we have tried to bring greater clarity into the bipartisan picture by working with Congressional leaders.

The Secretary commented that it is a good thing that the public is going to realize that the successful bipartisan handling of foreign policy problems is a difficult job. He added that the White House and the State Department will continue to have the ultimate responsibility in this field.

6. Presidential Speech on Foreign Policy

M r. Barrett indicated that the draft of this speech to be given at ASNE tomorrow will reach the Secretary today.

7. Congressional Attacks on State Department Security

While we develop the bipartisan approach, we will continue to defend ourselves no less vigorously along this front.

8. Senator Tydings' Suggestions to the Secretary

The Secretary mentioned that he always passes the Senator's suggestions on to appropriate people in the Department. Mr. Humelsine assured him that they were properly followed up.

9. Ambassador Muccio's Report on Korea

While the situation in Korea can not actually be considered satisfactory, the Koreans, with American help, have been attaining a degree of stability, particularly in the economic sphere. The joint economic committee has done some effective work. The inflationary cycle has been somewhat eased. Koreans have displayed courage in coming up with a balanced budget calling for more stringent taxes in spite of approaching elections. The military situation is the brightest. There is a trained Korean army which is ready to fight and which has been active in cleaning out guerrilla activities. Some fighting of this nature continues. The people enjoy a measurable degree of political independence and a freedom of the press, although not up to American

S E C R E T

- 4 -

13. **Arrival of Mr. Dulles**

It was explained that Mr. Dulles is arriving at the airport this morning, and that Mr. McWilliams has started to arrange appoint-ments with him, requested by people in the Department. Mr. Butterworth indicated that he would like to talk to him about the Japanese peace treaty.

14. **Attacks on Security**

Mr. Fisher asked for any helpful ideas for the Department's use in combatting the current charges from the Hill. It was felt that the only real danger to us will be that if the situation is unduly prolonged, the public may adopt a where there's smoke, there's fire attitude. It will, therefore, be all the more desirable to bring the affair to a close. It was felt that we might do something to show the injurious effect the situation is having abroad.

15. **Communist Infiltration in Foreign Affairs Groups**

The Department will give some study to the possibility that Communists may be infiltrating American foreign affairs organizations for the purpose of discrediting them. Mr. Armstrong will make an intelligence appraisal available to Mr. Fisher today.

cc: MR. McWilliams
     Mr. Sheppard
     Mr. Barnes
     Mr. Brown
     Mr. Wilgus
     Mr. Schwartz
     Mr. Sohm

MEMORANDUM FOR FILE

CONFIDENTIAL

Under Secretary's Meeting April 24, 1950

Public Relations on Security Issue

Mr. Barrett explained the events surrounding the Saturday night press "offensive" of the Secretary's. On Thursday night, Senator McCarthy spoke before the ASNE and apparently scored some success as a result of his manner if not of the substance of his remarks. Subsequently, a briefing session was set up in the Department which was well attended by editors, in which Mr. Webb and others presented matters of organization and substance. This briefing session was well received by the editors. On Saturday night before the ASNE, the Secretary followed up a prepared speech by about half an hour of off-the-record comments on the current attacks on the Department, which were extremely successful. However, members of the meeting were cautioned not to interpret that success as meaning that the current difficulties are over.

Report on the Fourth Session of the Contracting Parties of GATT

Mr. O'Gara introduced Mr. John Evans, ER, who served as Vice Chairman of the U.S. delegation at this five week session at Geneva under Ambassador Grady, who was Chairman. Other U.S. representatives included Agriculture, Treasury, Commerce, and the Office of the U.S. Special Representative. Mr. Evans explained that GATT is the multilateral agreement arising out of the bilaterals resulting from the Hull reciprocal trade agreement program. He stressed that GATT is just an agreement and not an organization. It differs from ITO in that it has a narrower membership than the ITO signing group at Havana and a narrower subject matter. It has no organization, no permanent secretariat, and no continuing governing body. It does have a convention, though a narrower one than the ITO's.

If the ITO Charter is ratified bringing the Organization into effect, GATT will disappear and its provisions will be adopted by the new organization.

GATT relations with OEEC are informal. The two supplement each other somewhat in Western Europe. GATT is a convention with specific rules of behavior. It can reduce tariff barriers, etc., which the OEEC cannot do. Yet, the OEEC has certain sanctions which the GATT does not. OEEC can reduce quantitative restrictions on trade more readily than GATT, which must proceed according to a formula.

Meeting Minutes under secretary's meeting May 29, 1950
RG 59 Entry A1-396C Minutes of under secretary's meetings UM Minutes Feb. 3, 1949-Jan. 25, 1952
Box 1

SECRET
-3-

Miscellaneous

Mr. Webb then mentioned briefly various important decisions made
in the absence of the Secretary and new programs launched, including
the loan to Argentina, the assignment of Mr. Dulles and Mr. Rusk to
work together on problems of the Far East, and Mr. Dulles' plans to visit
that area. He mentioned that the Philippines still present a problem
which he wanted to discuss with the Secretary.

Attacks on Department

Mr. Webb mentioned the continuing attacks on the Department which
appear increasingly to be directed at the whole government. Asked by
the Secretary whether the Department should continue to make public
statements concerning these attacks, Mr. Fisher said we must walk a
line between helping to keep the situation on the front page on the
one hand and permitting inaccuracies to go unrefuted on the other hand.
He felt that we owe it to the truth to respond, and that generally our
statements have had a favorable effect.

Dollar Gap

Mr. Thorp presented our 1949 trade figures which show exports in
the amount of $12 billion and imports of $6.6 billion, or a gap of
$5.4 billion. He added that 1949 is not too significant because of
the shift in exports which took place in the middle of the year as well
as the recession.

Compared to a year ago, our export rate has dropped almost
$4 billion. Import rates are up slightly, with a result that the
annual rate of the gap is now about $2 billion as compared to about
$6 billion a year ago. This narrowing results mostly from the drastic
reduction of exports. He noted that crude foodstuffs account for the
bulk of the drop. Raw cotton, on the other hand, accounts for the
largest item of increased export. The import situation is very uneven.
While there is a big increase in raw materials and foodstuffs, finished
manufactures are down about 7 per cent over a year ago. Coffee imports
have increased around 1/4 billion per annum while unmanufactured wool
shows the next heaviest import increase. Metals show the biggest drop
in imports.

Approaching the subject by area, he said that our exports have
held up best in North America. Except for Africa, where our volume
is small, the big cut-back has been in exports to Latin America,
mainly because of the import controls imposed there. Our imports

2205

ao 12

1   Mr. Humelsine. Very nearly an even split between the

2 two groups.

3   Senator Hoey. By "abroad" you mean somewhere in this

4 country or other countries?

5   Mr. Humelsine. In other countries.

6   Senator Hoey. Oh. A little over half of them were in

7 other countries?

8   Mr. Humelsine. Yes, sir, a little over half.

9   Senator Mundt. I believe Jack told me that number has

10 been expanded somewhat.

11   Mr. Humelsine. That number has. Since the 31st we have

12 gotten rid of 14 additional people, making a total of 105.

13   Senator Hoey. 105 total?

14   Mr. Humelsine. Yes, sir.

15   Senator Hoey. Have all of those been separated from the

16 service?

17   Mr. Humelsine. All of those have been separated.

18   Senator Hoey. And was the separation of each one, the

19 information on it, furnished to the Civil Service? You furn-

20 ished the information to the Civil Service on employees in

21 this country and then even gave the Civil Service the informa-

22 tion as to those in Foreign Service?

23   Mr. Humelsine. That is right.

24   Senator Mundt. To your knowledge, are there any people

25 in the State Department employed now who have a record of

2206

ao 13

1 homosexuality?

2 Mr. Humelsine. We have six cases under investigation.

3 Senator Mundt. When those are disposed of one way or an-

4 other, will that complete the whole list insofar as you are

5 aware?

6 Mr. Humelsine. No, sir. We have 20 that are on a basis

7 of allegations having been made against them.

8 Senator Mundt. That would be 26 that are still, you

9 might say, under suspicion?

10 Mr. Humelsine. Yes, sir. The allegations have been

11 made. It may turn out that they are false.

12 Senator Mundt. There are 20 in the category of allega-

13 tions made and still investigating. You have another category

14 of six I thought were allegations, too.

15 Mr. Humelsine. These are actually ones definitely under

16 investigation right at this moment. We have allegations on 20

17 others which we are just in process of starting. I mean, be-

18 fore you can actually come up and make that charge you have to

19 be pretty sure of your ground, and in those six cases we are

20 very sure of our ground.

21 Senator Mundt. Have you any information -- I may have

22 to get this from the Civil Service -- but you send them a fan-

23 fold, whatever that is?

24 Mr. Humelsine. That is just a personnel action to the

25 Civil Service Commission.

ao 25   2218

1   Mr. Hummelsine.   We do not handle under the loyalty and

2   security program this homosexuality problem. That is handled

3   administratively.

4   Mr. Flannagan.   Not under the security program at all?

5   Mr. Humelsine.   No. I want to get that straight. As

6   far as the security program I am talking about, the formal

7   security and loyalty program of the Department, we regard

8   homosexuals as a security risk but handle the administration

9   of them administratively. The reason we handle it adminis-

10   tratively, we find it a better way to eliminate those people

11   from the employment of the Government.

12   Mr. Flannagan.   What would be the difference between

13   handling them under the security program and handling them ad-

14   ministratively?

15   Mr. Humelsine.   Under the security program you would have

16   to go through this business of going through boards, and so

17   forth. Now, I do not want to intimate to you by this or sug-

18   gest to you that we are being unfair to the individuals by

19   handling it administratively, because we have found through

20   experience that about 95 per cent of these people voluntarily

21   confess when they are charged. We found that about 95 per cent

22   of them voluntarily confess. In fact, I think we have only one

23   case on record in all our handling of this problem in which a

24   person claimed not to be a homosexual.

25   Senator Smith.   Do many of them resign voluntarily?

2214-B

1   (2) The Department was requested to furnish the date on

2   which it adopted the procedure of notifying the Civil Service

3   Commission of the specific reasons for resignations.

4   The procedure was adopted on April 7, 1950. How-

5   ever, prior to that date the Department, when requested,

6   advised the Civil Service Commission as well as other

7   Government agencies of the real reason for resignations.

8   (3) The Department was requested to furnish the length

9   of time that the 105 individuals had been employed prior to

10 their resignation.

11   This information is as follows:

12   Less than 1 year 29

13   1 to 2 years   24

14   2 to 3 years   16

15   3 to 4 years   12

16   4 to 5 years   4

17   5 to 6 years   1

18   6 to 7 years   4

19   7 to 8 years   4

20   8 to 9 years   1

21   9 to 10 years 2

22   Over 10 years 8

23   Total 105

24

25 In analyzing these figures it should be borne in mind

    that the Department did not realize that it had a homosexual

May 2, 1951

My dear Mr. Collins:

I have your letter of April 28th, 1951, concerning the dismissal of homosexuals from the State Department's pay roll. I am glad to have the opportunity to discuss certain aspects of this problem with you.

First, I would like to assure you that I am as disturbed about the prevalance of homosexuality as you are. As I recently testified before Congress, I believe homosexuals are sick -- just as sick as people who have cancer. Nevertheless it is absolutely clear that homosexuals are very poor security risks and as such have no place in the Department of State or in the Foreign Service. We do not want them on our pay rolls and for several years we have carried out a vigorous program to separate them. As proof of our efforts to this end, I informed Congress that 144 homosexual security risks had been retired from the Department's rolls since 1947. This process of ferreting out undesirables is a continuing one and you may be assured that whenever we receive adverse information about one of our employees, the matter is immediately investigated.

You may be interested to know that the Department participates fully in the President's loyalty program, which is discussed in some detail in the enclosed pamphlet entitled "Loyalty and Security in the Department of State". In addition, the State Department carries out rigid regulations and procedures of its own to make sure that only responsible and loyal persons hold positions in our service.

Mr. C. F. Collins,
50 Lancaster Street,
Leominster, Massachusetts.

- 2 -

I have a security staff of 104 members, headed by a former agent of the F.B.I. Under him are 74 trained investigators who carefully check employees in our continuing security program and who make a thorough investigation of the background of all prospective new employees. The investigations these men make are complete and cover every aspect of the applicant's character. I might add that in addition to looking into an applicant's or an employee's background, neighborhood, and associates, we ascertain in a discreet way whether there is any indication of homosexuality. Findings of these investigators are studied by expert evaluators and questionable applicants are rejected outright.

With respect to your question as to the over all percentage of homosexuals in the United States, I have discussed this matter with several medical specialists in this field. Their most consistent estimate is that homosexuality runs about 4 per cent of the total population. Considering that the Department today has 26,000 employees at home and abroad, and including aliens employed overseas, who are also subject to regulation investigations, the proportion of dismissals for perversion in our service is not high.

The officers and employees of the State Department and Foreign Service have as their greatest concern the welfare of the United States. Their record in war and peace testifies to their loyalty and devotion to this government. You may be interested to know that the Department of State has the second highest percentage of veterans of all the departments and agencies of this government. Approximately 80 per cent of our men are veterans, and of these nearly 300 are disabled.

I hope that these facts will be of interest to you and that you will have the opportunity to read the enclosed pamphlet. With all good wishes,

Sincerely,

Carlisle H. Humelsine
Deputy Under Secretary

Enclosure:

Pamphlet: "Loyalty and
Security in the Department
of State"

PA:PL:PCrane:PHCulley:vmq   PL:JMP   PER   SY   5/2/51

NOV   1951

In reply refer to
SY

My dear Mr. Javits:

    Reference is made to your request of October 12, 1951 for
information which you might furnish to a constituent concerning
the Loyalty and Security Program of the Department of State.

    Since your correspondent specifically links "Loyalty and
Security Risk investigations," I am sending combined figures
on the program. You understand, of course, that loyalty is
adjudicated under the President's Loyalty Program, whereas
security is a matter of departmental concern alone under
statutory authority.

    Although your constituent mentions no specific data, it is
believed that he will be interested in the following information
regarding the Department's security program since January 1947.
It should be noted that the President's Loyalty Program was
inaugurated March 21, 1947. State Department records disclose
that from January 1, 1947 to August 15, 1951, 347 individuals
on whom some security question existed left the Department. This
figure includes individuals who left the Department by resignation
or for other reasons. It should be noted to your constituent that
the 347 persons were not all proven security risks and that if the
investigations in their cases had been completed, many of the
security questions would probably have been resolved favorably.

    The Department has found sixteen employees to be security
risks since January 1947. These employees have been separated
from the Department.

        Sincerely yours,

        For the Acting Secretary

        Carlisle H. Humelsine
        Deputy Under Secretary

The Honorable
    Jacob K. Javits,
        House of Representatives.
CON:SY:WDTemmay:mse

Correspondence to embassies March 29, 1951
RG 59-P-528 General Records of the Department of State, Office of the Legal Advisor, Records Relating to
Loyalty and Security Issues 1944-1954
Box 1 File-Security Program-Humelsine

DEPUTY UNDER SECRETARY OF STATE

WASHINGTON

PERSONAL AND CONFIDENTIAL   March 29, 1951

My dear Mr.

    I am enclosing a report entitled "Employment of Homosexuals
and Other Sex Perverts in Government". This report was submitted
on December 15, 1950, by the Subcommittee on Investigations, Committee on Expenditures in the Executive Department.

    As you can see by a study of this report, the problem of
employment of sex perverts is one of considerable concern to the
Government. My interest in calling this matter to your attention
is to emphasize that the Department and the Foreign Service must
take all possible action to prevent the employment of perverts.

    It is with reluctance that I call to your attention this
unpleasant problem, but it is also because of this very reluctance
that I must do so. It is entirely natural, of course, for normal
individuals to shy away from the question of perversion. The danger
is that this natural inhibition may lead us to close our eyes to
the problem rather than to be alert to it. We must not permit ourselves to succumb to this potential danger.

    The enclosed report shows that 91 homosexuals were separated
from the Department and Foreign Service rolls between January 1947
and January 1950. The report also points out that the Department
of State considers sex perverts to be security risks. The Department's view in this regard is certainly not new to you, but it
should be made clear, should the question arise, that the Department's attitude is not based on arbitrary assumptions. Perverts
are considered security risks because there is ample evidence in
the Government to justify this opinion.

    Here in Washington we are exerting every effort to prevent
employment of perverts or persons having a tendency toward perversion. I am even more concerned about the problem of perversion in

                                                        the

PERSONAL AND CONFIDENTIAL

Correspondence to embassies March 29, 1951 RG 59-P-528 General Records of the Department of
State, Office of the Legal Advisor, Records Relating to Loyalty and Security Issues 1944-1954
Box 1 File-Security Program-Humelsine

PERSONAL AND CONFIDENTIAL

- 2 -

the field, not only as it applies to Foreign Service personnel,
but as it may apply to non-American and local employees. I say
that I am more concerned about the field situation because I rec-
ognize that greater facilities are available here for ferreting
out such individuals.

I must ask you to personally examine the personnel situation
at your post and to take any steps necessary to assure yourself
that no American or locally employed personnel under your jurisdic-
tion are sex perverts. The Division of Foreign Service Personnel,
the Division of Security and the Foreign Service Inspectors have
been directed to give you any assistance you may require. One of
these three organizations should be notified immediately upon
evidence or suspicion of sex perversion on the part of any person
assigned to your post. The Department should also be notified of
evidence or suspicion of perversion in regard to Americans formerly
employed by the Department or the Foreign Service.

I assure you that any information you provide will be handled
with the utmost discretion and that a thorough and impartial investigation will be made so that appropriate action can be taken.

I should appreciate your full cooperation in this unpleasant
matter.

Sincerely yours,

Carlisle H. Humelsine

Enclosures.

  1. Press Release No. 233
  2. Senate Document No. 241

PERSONAL AND CONFIDENTIAL